W9-ATB-449

Everyday Mathematics®

The University of Chicago School Mathematics Project

Student Math Journal
Volume 1

Grade 4

McGraw Hill Wright Group

The McGraw-Hill Companies

The University of Chicago School Mathematics Project (UCSMP)
Max Bell, Director, UCSMP Elementary Materials Component; Director, *Everyday Mathematics* First Edition
James McBride, Director, *Everyday Mathematics* Second Edition
Andy Isaacs, Director, *Everyday Mathematics* Third Edition
Amy Dillard, Associate Director, *Everyday Mathematics* Third Edition

Authors

Max Bell	Andy Isaacs	Robert Balfanz*
John Bretzlauf	James McBride	William Carroll*
Amy Dillard	Kathleen Pitvorec	Sheila Sconiers*
Robert Hartfield	Peter Saecker	

**First Edition only*

Technical Art
Diana Barrie

Photo Credits
© Gregory Adams/Getty Images, cover, *top right;* Getty Images, cover, *center;* ©Tony Hamblin; Frank Lane Picture Agency/CORBIS, cover, *bottom left;* ©Nick Rowe/Getty Images, p. v.

Teachers in Residence
Carla L. LaRochelle, Rebecca W. Maxcy

Editorial Assistant
Laurie K. Thrasher

Contributors
Martha Ayala, Virginia J. Bates, Randee Blair, Donna R. Clay, Vanessa Day, Jean Faszholz, James Flanders, Patti Haney, Margaret Phillips Holm, Nancy Kay Hubert, Sybil Johnson, Judith Kiehm, Deborah Arron Leslie, Laura Ann Luczak, Mary O'Boyle, William D. Pattison, Beverly Pilchman, Denise Porter, Judith Ann Robb, Mary Seymour, Laura A. Sunseri-Driscoll

This material is based upon work supported by the National Science Foundation under Grant No. ESI-9252984. Any opinions, findings, conclusions, or recommendations expressed in this material are those of the authors and do not necessarily reflect the views of the National Science Foundation.

www.WrightGroup.com

Wright Group

Copyright © 2007 by Wright Group/McGraw-Hill.

All rights reserved. Except as permitted under the United States Copyright Act, no part of this publication may be reproduced or distributed in any form or by any means, or stored in a database or retrieval system, without the prior written permission from the publisher, unless otherwise indicated.

Printed in the United States of America.

Send all inquiries to:
Wright Group/McGraw-Hill
P.O. Box 812960
Chicago, IL 60681

ISBN 0-07-604582-X

1 2 3 4 5 6 7 8 9 QWD 12 11 10 09 08 07 06

The McGraw-Hill Companies

Contents

UNIT 1 Naming and Constructing Geometric Figures

Welcome to *Fourth Grade Everyday Mathematics* .1
Using Your *Student Reference Book* 2
Math Boxes 1◆1 3
Points, Line Segments, Lines, and Rays 4
Math Boxes 1◆2 5
Angles 6
Addition and Subtraction 7
Math Boxes 1◆3 8
Math Boxes 1◆4 9
Parallelograms 10
What Is a Polygon? 12
Math Boxes 1◆5 13
An Inscribed Square 14
Math Boxes 1◆6 16
Circle Constructions 17
Math Boxes 1◆7 20
Copying a Line Segment 21
Hexagons in Our World 22
Constructing an Inscribed, Regular Hexagon 23
More Constructions 24
Definition Match 25
Math Boxes 1◆8 26
Math Boxes 1◆9 27

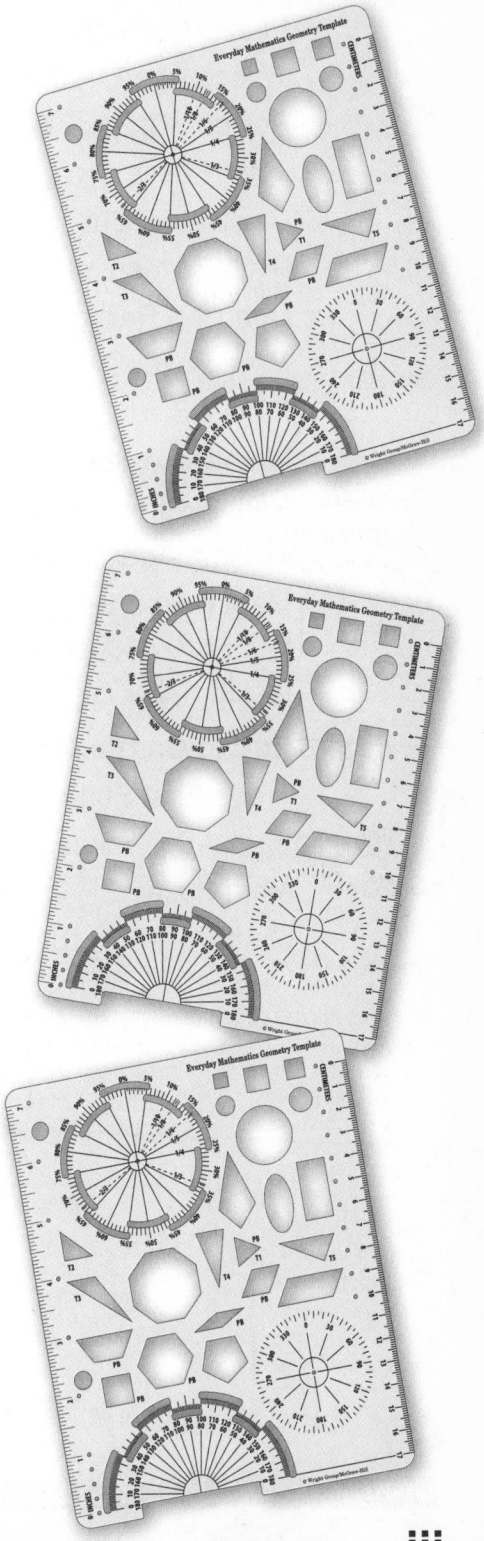

UNIT 2 Using Numbers and Organizing Data

A Visit to Washington, D.C. 28
Math Boxes 2•1 29
Name-Collection Boxes 30
Math Boxes 2•2 31
Place-Value Chart 32
Taking Apart, Putting Together 33
Polygon Checklist 34
Math Boxes 2•3 35
Calculator "Change" Problems 36
Math Boxes 2•4 37
Counting Raisins 38
Math Boxes 2•5 39
Family Size 40
Math Boxes 2•6 41
Partial-Sums Addition 42
Column Addition 43
Math Boxes 2•7 44
Math Boxes 2•8 45
Head Sizes 46
How to Construct a Kite 48
Trade-First Subtraction 49
Partial-Differences Subtraction 50
Math Boxes 2•9 51
Math Boxes 2•10 52

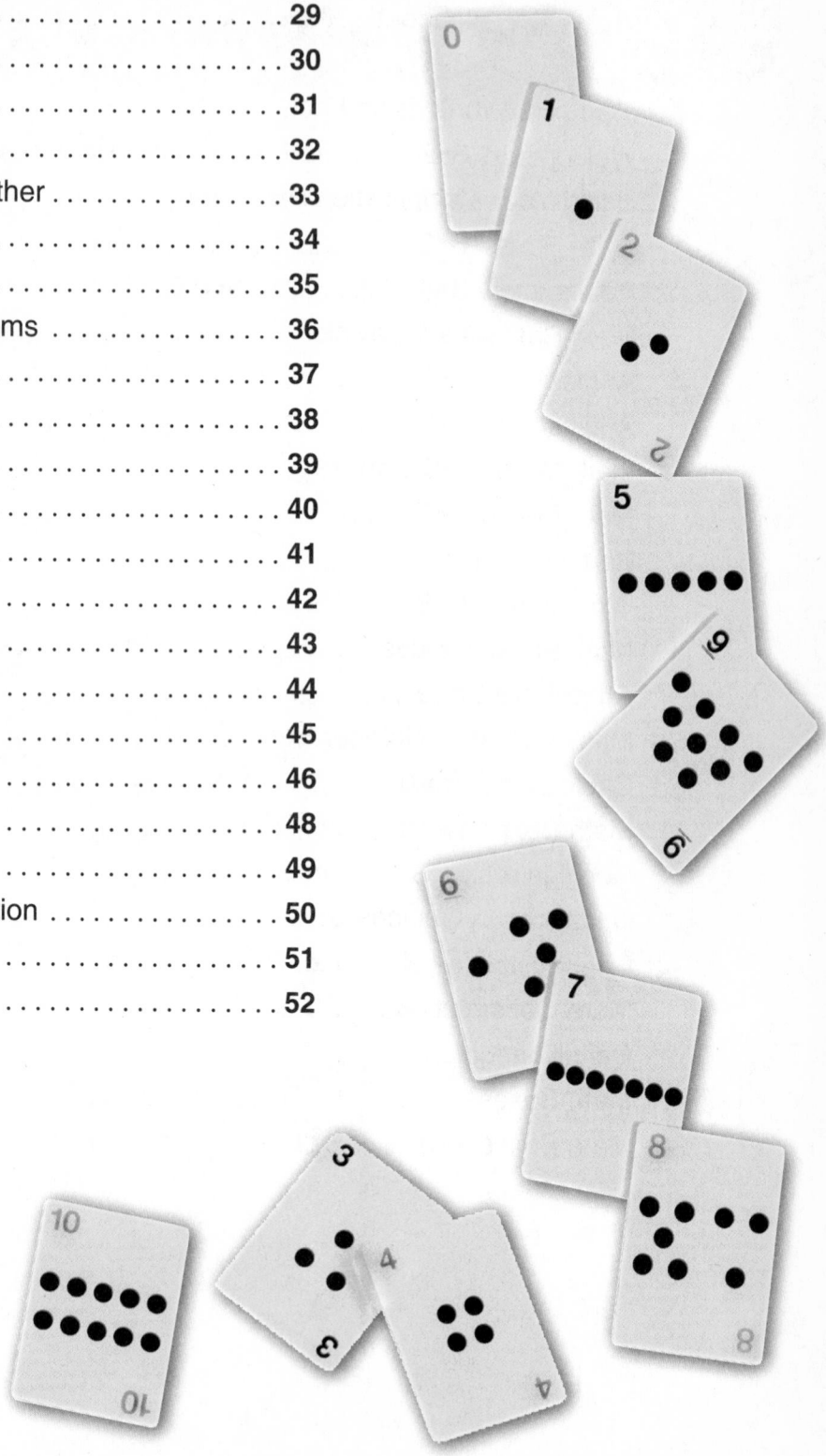

UNIT 3 Multiplication & Division: Number Sentences & Algebra

"What's My Rule?" 53
A Polygon Alphabet 54
Math Boxes 3•1 55
Multiplication/Division Facts Table 56
Math Boxes 3•2 57
Patterns in Multiplication Facts 58
Math Boxes 3•3 59
Math Boxes 3•4 60
Multiplication and Division 61
Math Boxes 3•5 62
Math Boxes 3•6 63
Measuring Air Distances 64
Math Boxes 3•7 65
Flying to Cairo 66
Math Boxes 3•8 68
Number Sentences 69
Math Boxes 3•9 70
Parentheses in Number Sentences 71
Math Boxes 3•10 72
Broken Calculator 73
Open Sentences 74
Estimating Distances 75
Math Boxes 3•11 76
Math Boxes 3•12 77

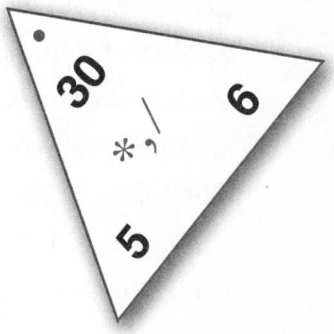

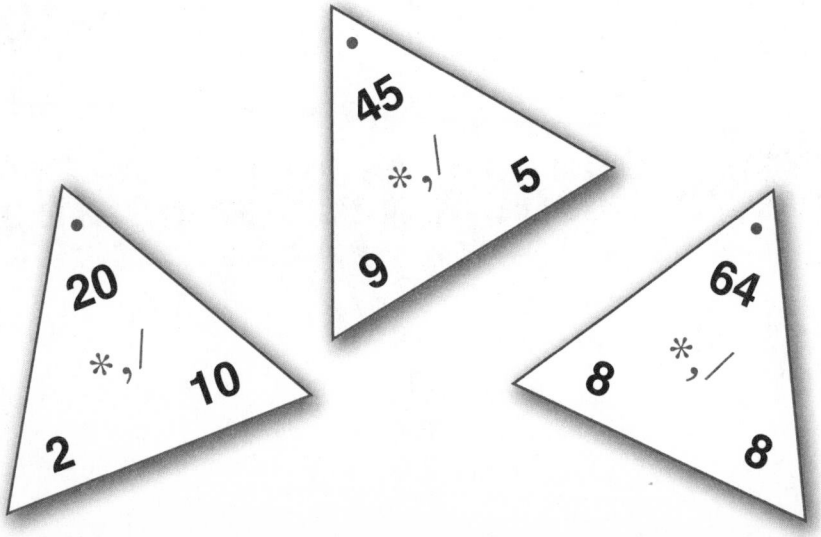

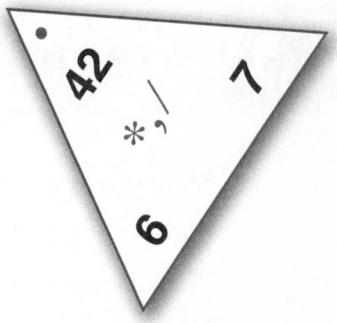

UNIT 4 Decimals and Their Uses

Place-Value Number Lines 78
Math Boxes 4•1 79
Tenths and Hundredths 80
Math Boxes 4•2 81
Comparing Decimals 82
Ordering Decimals 83
Math Boxes 4•3 84
A Bicycle Trip 85
Math Boxes 4•4 86
Decimal Addition and Subtraction 87
Circle Graphs 88
Math Boxes 4•5 89
Keeping a Bank Balance 90
Math Boxes 4•6 92
Math Boxes 4•7 93
Tenths, Hundredths, and Thousandths 94
Measuring Length with Metric Units 96
Math Boxes 4•8 97
Personal References for Units of Length 98
Measurement Collection for
Metric Units of Length 99
Math Boxes 4•9 100
Measuring in Millimeters 101
Measuring Land Invertebrates 102
Math Boxes 4•10 104
Math Boxes 4•11 105

UNIT 5 Big Numbers, Estimation, and Computation

Multiplying Ones by Tens 106
Multiplying Tens by Tens 107
Math Boxes 5•1 108

Presidential Information 109
Math Boxes 5•2 110
Math Boxes 5•3 111
Estimated U.S. Distances and Driving Times . . . 112
Planning a Driving Trip 113
What Do Americans Eat? 114
Estimating Averages 115
Math Boxes 5•4 116
Math Boxes 5•5 117
The Partial-Products Method 118
My Measurement Collection for Units of Length . . 120
Math Boxes 5•6 121
Multiplication Number Stories 122
Lattice Multiplication 124
Math Boxes 5•7 125
Reading and Writing Big Numbers 126
How Much Are a Million and a Billion? 127
Internet Users 128
Math Boxes 5•8 129
Place Value and Powers of 10 130
Math Boxes 5•9 131
Evaluating Large Numbers 132
Math Boxes 5•10 133
Traveling to Europe 134
Math Boxes 5•11 135
Math Boxes 5•12 136

UNIT 6 Division; Map Reference Frames; Measures of Angles

Math Boxes 6•1 137
Multiplication/Division Number Stories 138
Extended Multiplication Facts 140

Math Boxes 6•2 141
Solving Division Problems 142
Partial-Quotients Division Algorithm 144
Place Value in Decimals 146
Math Boxes 6•3 147
Interpreting Remainders 148
Math Boxes 6•4 150
Math Boxes 6•5 151
Making a Full-Circle Protractor 152
Clock Angles 153
Population Bar Graph 154
Measuring Angles 155
Math Boxes 6•6 156
Drawing and Measuring Angles 157
Drawing Angles 158
Math Boxes 6•7 159
Math Boxes 6•8 160
Map of the Island of Ireland 161
A Campground Map 162
Finding Distances on a Map 163
Locating Places on Regional Maps 164
Math Boxes 6•9 165
Partial-Quotients Division 166
Math Boxes 6•10 168
Math Boxes 6•11 169
My Route Log 171
Route Map 172
My Country Notes 174

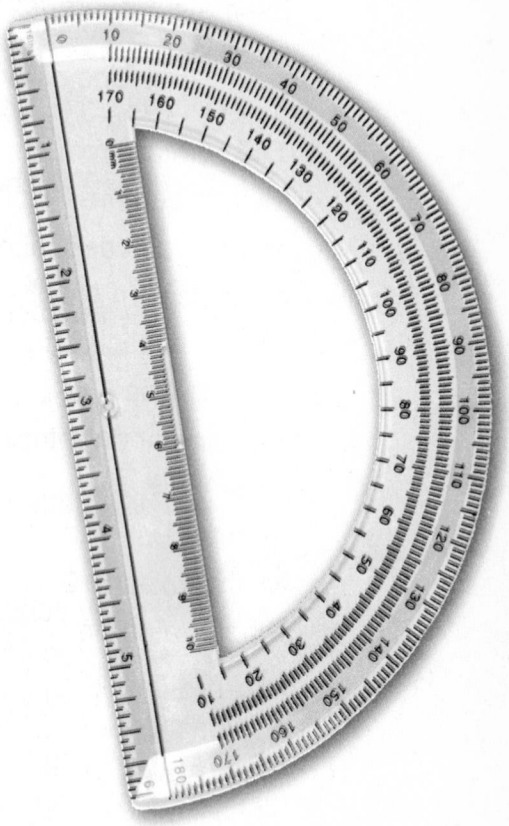

Activity Sheets

*, / Fact Triangles 1 . . . . . **Activity Sheet 1**
*, / Fact Triangles 2 . . . . . **Activity Sheet 2**
*, / Fact Triangles 3 . . . . . **Activity Sheet 3**
*, / Fact Triangles 4 . . . . . **Activity Sheet 4**

LESSON 1•1 Welcome to *Fourth Grade Everyday Mathematics*®

Much of your work in Kindergarten through third grade was basic training in mathematics and its uses. You learned to solve number stories and use arithmetic, including basic addition and multiplication facts.

Fourth Grade Everyday Mathematics builds on this basic training and begins to make the transition to mathematics concepts and ways of using mathematics that are more like what your parents and older siblings may have learned in high school. The authors believe that fourth graders can do more than was thought possible in years past.

Here are some things you will be asked to do in *Fourth Grade Everyday Mathematics:*

- Increase your "number sense," "measure sense," and estimation skills.
- Extend your skills in the basics of arithmetic—addition, subtraction, multiplication, and division. There is not much more to learn about the arithmetic of whole numbers, but over the next couple of years, you will become comfortable using fractions, percents, and decimals.
- Learn about variables (letters that stand for numbers) and other introductory topics in algebra.
- Develop your geometry concepts and skills with more exact definitions and classification of geometric figures, constructions and transformations of figures, and investigation of areas and volumes of shapes.
- Take a World Tour. Along the way, you will consider many kinds of data about various countries and learn how to use coordinate systems to locate places on world globes and maps.
- Do many projects involving numerical data.

In fourth grade, you will be asked to do more independent reading and investigation (often working with partners or in groups) rather than being told everything by your teacher.

We hope that you find the activities fun and that you see the beauty in mathematics. Most importantly, we hope you become better and better at using mathematics to solve interesting problems in your life.

Date Time

LESSON 1•1

Using Your *Student Reference Book*

Use your *Student Reference Book* to complete the following:

1. Look up the word **mode** in the Glossary.

   a. Copy the definition. _____

   _____

   b. On which page in the *Student Reference Book* could you find more information about the mode of a set of data? page _____

2. Find the essay "Comparing Numbers and Amounts."

   a. Describe what you did to find the essay.

   _____

   _____

   b. Read the essay and solve the Check Your Understanding problems.

   Problem 1: _____ Problem 2: _____

   Problem 3: _____ Problem 4: _____

   c. Check your answers using the Answer Key.

3. Look up the rules for the game *Name That Number.*

   a. On which page did you find the rules? page _____

   b. How many players are needed for the game? _____ players

4. Go to the World Tour section. Record two interesting facts you find there.

   a. Fact 1: _____

   I found this information on page _____.

   b. Fact 2: _____

   I found this information on page _____.

Date Time

LESSON 1•1

Math Boxes

1. Add mentally.

a. 1 + 7 = _____

b. 4 + 0 = _____

c. 5 + 5 = _____

d. 2 + 9 = _____

e. 8 + 5 = _____

f. 7 + 9 = _____

2. Fill in the missing numbers and state the rule.

a. 2, 4, 6, 8, _____, _____

Rule: +2

b. 65, 60, 55, _____, _____, _____

Rule: _____

c. 109, 95, 81, _____, _____, _____

Rule: _____

SRB 160 161

3. Complete.

21 in. = _____ ft _____ in.

Circle the best answer.

A. 1 ft 1 in.

B. 1 ft 10 in.

C. 1 ft 9 in.

D. 1 ft 3 in.

SRB 129

4. Complete.

a. 2 quarters = _____ dimes

b. 1 dollar
and 5 nickels = _____ quarters

c. 14 dimes = _____ pennies

d. 8 quarters = _____ dollars

e. 3 quarters
and 9 nickels = _____ dimes

5. Add mentally or with a paper-and-pencil algorithm.

a. 32 + 35 = _____

b. 38 + 66 = _____

SRB 10 11

Date Time

LESSON 1•2 Points, Line Segments, Lines, and Rays

Use a straightedge to draw the following:

1. a. Draw and label line segment *RT* ($\overline{RT}$).

b. What is another name for $\overline{RT}$? _____

2. a. Draw and label line *BN* ($\overleftrightarrow{BN}$). Draw and label point *T* on it.

b. What are 2 other names for $\overleftrightarrow{BN}$? _____

3. a. Draw and label ray *SL* ($\overrightarrow{SL}$). Draw and label point *R* on it.

b. What is another name for $\overrightarrow{SL}$? _____

4. a. Draw a line segment from each point to each of the other points.

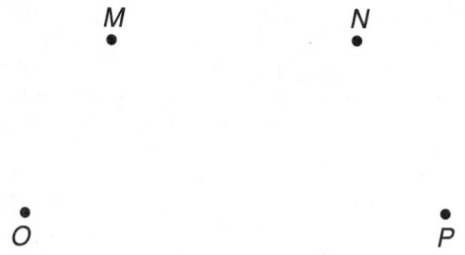

b. How many line segments did you draw? _____

c. Write a name for each line segment you drew.

_____

Date Time

LESSON 1·2

Math Boxes

1. Subtract mentally.

   a. 6 − 3 = _____

   b. 7 − 4 = _____

   c. 14 − 7 = _____

   d. 16 − 9 = _____

   e. _____ = 9 − 4

   f. _____ = 17 − 9

2. Draw and label line *QR*.
   Draw point *S* on it.

   What are two other names for line *QR*?

   _____

   _____

SRB 91

3. Complete.

   Max read _____ books.

   Sue read _____ books.

   Ira read _____ books.

   Pat read _____ books.

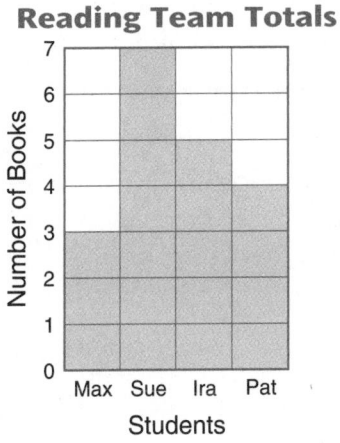

SRB 76

4. Cross out the names that do not belong in the name-collection box. Label the box with the correct number.

25 − 13
20 − 7
6 × 2
4 × 3
40 − 23
7 × 3

SRB 149

5. Subtract mentally or with a paper-and-pencil algorithm.

   a. 86 − 21 = _____

   b. 93 − 24 = _____

SRB 12–15

Date Time

LESSON 1•3 Angles

1. Which angle is bigger,

   $\angle ABC$ or $\angle DEF$? _____

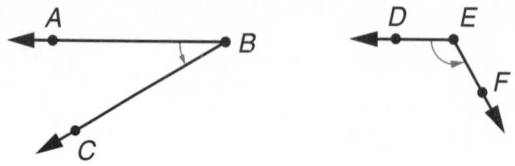

2. Draw $\angle BAC$. What is another name for $\angle BAC$? _____

3. What is the vertex of $\angle BAC$? Point _____

4. Feng said the name of this angle is $\angle SRT$. Is he right? Explain.

   _____

   _____

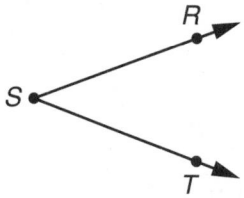

Try This

Use the points shown on the grid below and a straightedge to draw the angles.

5. Draw $\angle AED$.
   a. What is the vertex of the angle? Point _____
   b. What is another name for $\angle AED$? $\angle$ _____

6. Draw a right angle whose vertex is point C. My angle is called $\angle$ _____.

7. Draw an angle that is smaller than a right angle. My angle is called $\angle$ _____.

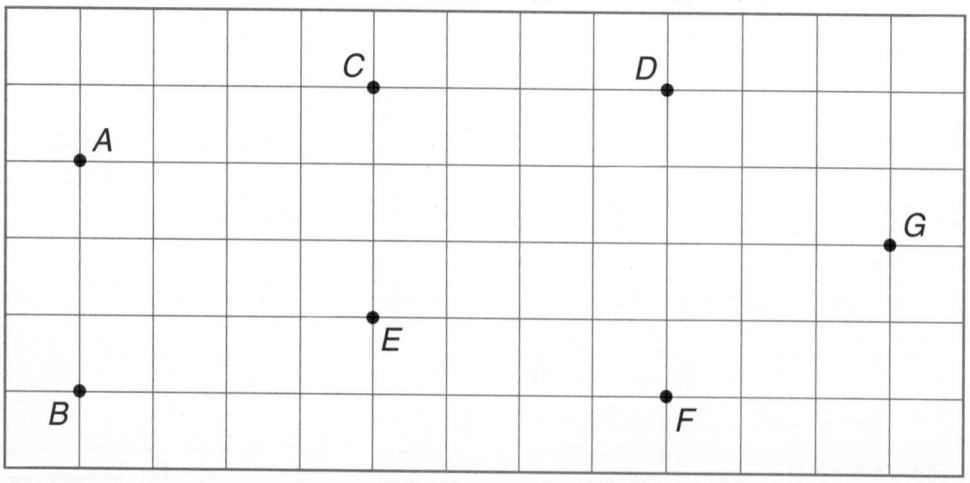

Date Time

LESSON 1•3

Addition and Subtraction

Use your favorite addition and subtraction methods to solve the problems. Show your work.

SRB 10–15

1. 48 + 79 = _____
2. 392 + 75 = _____
3. _____ = 296 + 514
4. _____ = 937 + 652
5. _____ = 98 − 57
6. 122 − 82 = _____
7. _____ = 900 − 632
8. _____ = 512 − 239

Date Time

LESSON 1•3

Math Boxes

1. Add mentally.

   a. 2 + 5 = _____

   b. 3 + 3 = _____

   c. 4 + 6 = _____

   d. _____ = 6 + 7

   e. _____ = 9 + 8

   f. _____ = 4 + 7

2. Fill in the missing numbers and state the rule.

   a. 4, 8, 12, 16, _____, _____, _____

   Rule: _____

   b. 33, 30, 27, _____, _____, _____

   Rule: _____

   c. _____, _____, _____, 106, 141, 176

   Rule: _____

SRB 160 161

3. Complete.

   a. 1 ft = _____ in.

   b. 24 in. = _____ ft

   c. _____ yd = 36 in.

   d. 30 in. = _____ ft _____ in.

   e. 50 in. = _____ yd _____ ft _____ in.

SRB 129

4. Complete.

   a. 9 dimes = _____ pennies

   b. 30 dimes = _____ dollars

   c. 4 quarters = _____ dimes

   d. 2 dollars and 10 nickels = _____ quarters

   e. 13 dollars = _____ quarters

5. Add mentally or with a paper-and-pencil algorithm.

   a. 63 + 12 = _____

   b. 56 + 97 = _____

SRB 10 11

Date Time

LESSON 1•4

Math Boxes

1. Subtract mentally.

a. 10 − 4 = _____

b. _____ = 8 − 5

c. 7 − 4 = _____

d. 15 − 7 = _____

e. 13 − 8 = _____

f. _____ = 17 − 8

2. Draw and label line *AB*.
Draw point *C* on it.

What are two other names for line *AB*?

_____

_____

SRB 91

3. Complete.

Luz sold _____ boxes.

Ana sold _____ boxes.

Mya sold _____ boxes.

Pei sold _____ boxes.

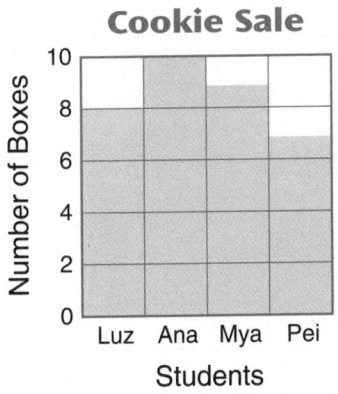

SRB 76

4. Which of these can go in a name-collection box for the number 50? Circle the best answer.

A. 10 + 35

B. 136 − 51

C. 200 ÷ 4

D. 4 × 15

SRB 149

5. Subtract mentally or with a paper-and-pencil algorithm.

a. 76 − 41 = _____

b. 52 − 38 = _____

SRB 12–15

Date Time

LESSON 1•4 Parallelograms

1. Circle the pairs of line segments below that are parallel. Check some of your answers by extending each pair of segments to see if the two segments in the pair meet or cross.

a.

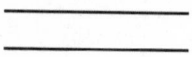

b.

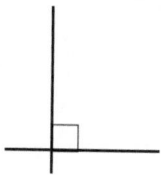

c.

d.

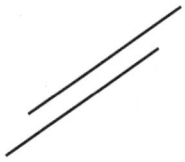

e.

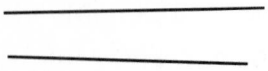

f.

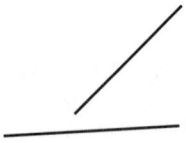

Use your Geometry Template or straightedge to draw the following quadrangles:

2. Draw a quadrangle that has 2 pairs of parallel sides.

This is called a _____.

3. Draw a quadrangle that has only 1 pair of parallel sides.

This is called a _____.

LESSON 1•4

Parallelograms *continued*

For Problems 4 and 5, circle the best answer(s). Some items have more than 1 correct answer, so you may need to circle more than 1 answer.

4. A **parallelogram** is a quadrangle that has 2 pairs of parallel sides. Which are parallelograms?

   A. squares

   B. rectangles

   C. rhombuses

   D. trapezoids

5. A **rhombus** is a parallelogram in which all sides are the same length. Which are always rhombuses?

   A. squares

   B. rectangles

   C. trapezoids

   D. kites

Try This

A **rectangle** is a parallelogram that has all right angles. Which of the following are rectangles? Write *always, sometimes,* or *never* to complete each sentence. Explain your answers.

6. Squares are _____ rectangles. Explain. _____

   _____

7. Rhombuses are _____ rectangles. Explain. _____

   _____

8. Trapezoids are _____ rectangles. Explain. _____

   _____

9. A kite is _____ a parallelogram. Explain. _____

   _____

Date Time

LESSON 1•5

What Is a Polygon?

These are polygons.

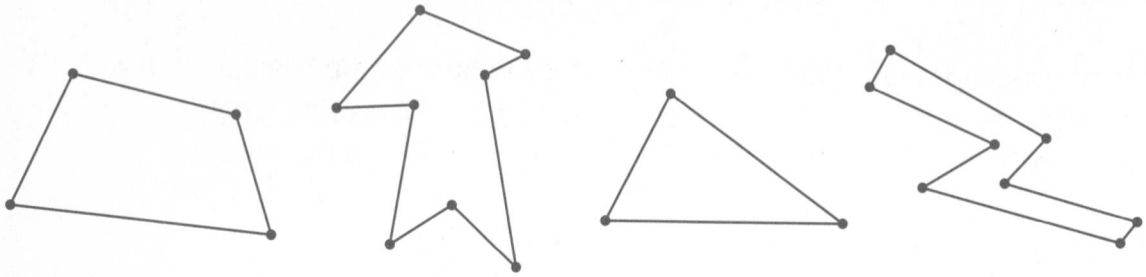

1 2 3 4

These are NOT polygons.

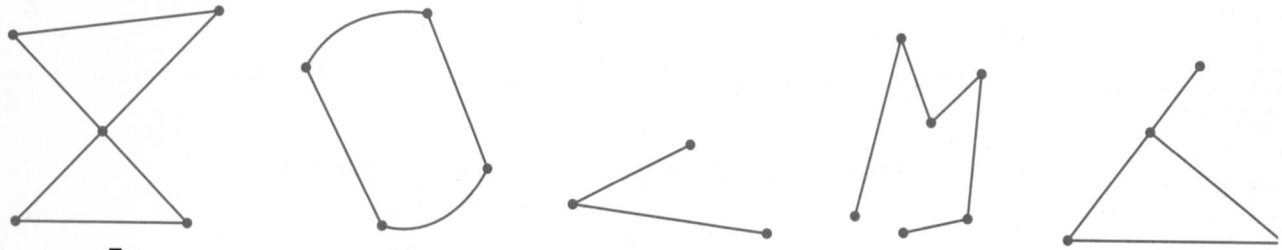

5 6 7 8 9

1. If you had to explain what a polygon is, what would you say? (*Think:* What do Polygons 1–4 have in common? How are Shapes 5–9 different from Polygons 1–4?)

_____

_____

_____

_____

_____

2. Choose one of the shapes from above. Explain why the shape is not a polygon.

_____

_____

_____

_____

_____

Date Time

LESSON 1•5

Math Boxes

1. Subtract mentally.

a. 7 − 0 = _____

b. 10 − 7 = _____

c. _____ = 9 − 4

d. _____ = 14 − 6

e. 13 − 7 = _____

f. 16 − 9 = _____

2. Draw ∠*MRT*.

What is another name for ∠*MRT*?

_____

SRB 92

3. Draw and label line segment *AB*.

What is another name for $\overline{AB}$?

_____

SRB 90

4. Name as many rays as you can in the figure below.

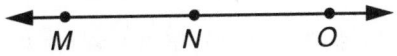

Write their names.

_____

_____

_____

SRB 91

5. Which polygons have 2 pairs of parallel sides? Circle the best answer.

A. square and trapezoid

B. rectangle and rhombus

C. triangle and parallelogram

D. pentagon and square

SRB 99 100

6. Put these numbers in order from least to greatest.

10,005 51,000

5,100 10,500

_____

_____

_____

_____

SRB 4

Date Time

LESSON 1•6

An Inscribed Square

SRB 116

Follow the directions below to make a square that you will tape on the next page.

Step 1 Use your compass to draw a circle on a sheet of colored paper. The circle should be small enough to fit on the next page. Cut out the circle.

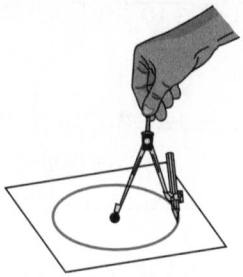

Step 2 With your pencil, make a dot in the center of the circle, where the hole is, on both the front and the back.

Step 3 Fold the circle in half. Make sure that the edges match and that the fold line passes through the center. Be sure to make sharp creases.

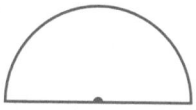

Step 4 Fold the folded circle in half again so that the edges match.

Step 5 Unfold your circle. The folds should pass through the center of the circle and form 4 right angles.

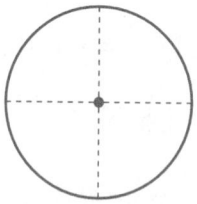

Step 6 Using a straightedge, connect the endpoints of the folds at the edge of the circle to make a square. Cut out the square.

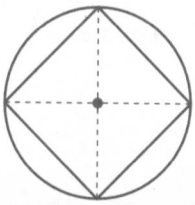

LESSON 1•6

An Inscribed Square *continued*

Now use your compass to find out whether the 4 sides of your square are about the same length.

Place the anchor on one endpoint of a side and the pencil point on the other endpoint of the side. Then, *without changing the compass opening,* try to place the anchor and pencil point on the endpoints of each of the other sides.

If the sides of your square are about the same length, tape the square in the space below. If not, follow the directions on page 14 again. Tape your best square in the space below.

Date Time

LESSON 1•6

Math Boxes

1. Subtract mentally.

   a. 9 − 5 = _____

   b. 11 − 2 = _____

   c. _____ = 14 − 7

   d. _____ = 12 − 4

   e. 13 − 6 = _____

   f. _____ = 12 − 9

2. Which of the shape(s) below are NOT polygons? _____

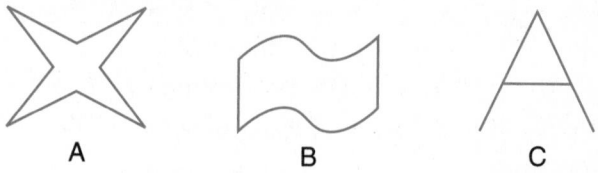

SRB 96

3. Draw a quadrangle with only 1 right angle. Draw in the right angle symbol.

   How do you know it is a right angle?

   _____

SRB 93 99

4. Circle the convex polygon(s).

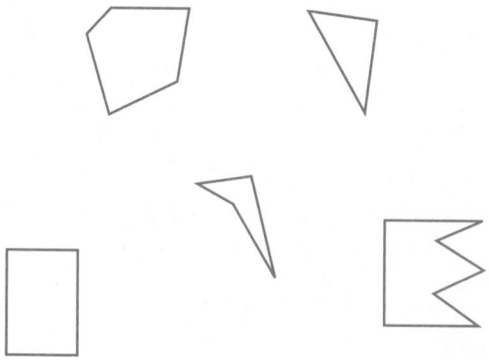

SRB 97

5. Draw and label ray *HA*. Draw point *T* on it.

   What is another name for $\overrightarrow{HA}$? _____

SRB 91

6. In the numeral 42,318, the 2 stands for 2,000.

   a. The 1 stands for _____.

   b. The 8 stands for _____.

   c. The 4 stands for _____.

   d. The 3 stands for _____.

SRB 4

Date Time

LESSON 1•7 Circle Constructions

Do each of the following 3 constructions on a separate sheet of paper. Try and try again until you are satisfied with your work. Then cut out your 3 best constructions and tape them in your journal.

1. Use your compass to draw a picture of a circular dartboard. It is not necessary to include the details of the board. Tape your best work in the space below. The circles in the dartboard and in your picture are called **concentric circles.**

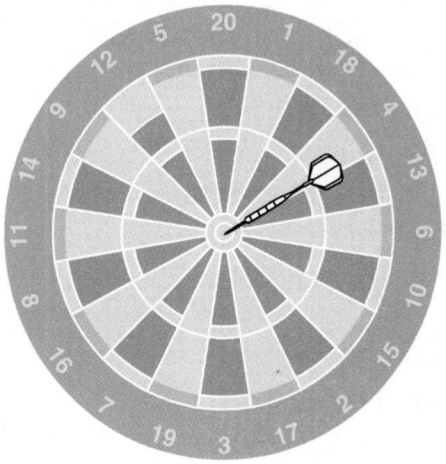

LESSON 1•7

Circle Constructions *continued*

2. a. Make a dot near the center of your paper. Use your compass to draw a circle with that dot as its center.

b. *Without changing the opening of your compass,* draw a **congruent** circle that **intersects** the center of the first circle. Mark the center of the second circle.

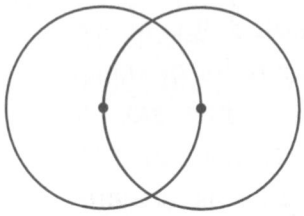

c. *Without changing the opening of your compass,* draw a third congruent circle that intersects the center of each of the first 2 circles.

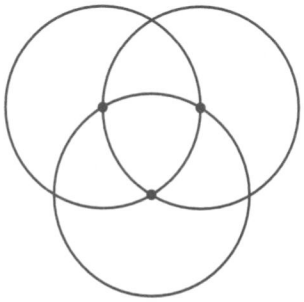

Try and try again until you are satisfied with your work. Then cut out your circle design and tape it in the space below.

LESSON 1•7

Circle Constructions *continued*

Try This

3. Draw this design with your compass. Work on separate sheets of paper until you are satisfied with your work. Color your best design. Then cut it out and tape it in the space below.

   Hint: Start by making the 3-circle design on page 18. Then add more circles to it.

Date Time

LESSON 1•7

Math Boxes

1. Subtract mentally.

a. 9 − 7 = _____

b. 10 − 6 = _____

c. _____ = 16 − 8

d. _____ = 17 − 7

e. 13 − 7 = _____

f. _____ = 15 − 9

2. Draw ∠*TIF*. What is the vertex of ∠*TIF*?

Point _____

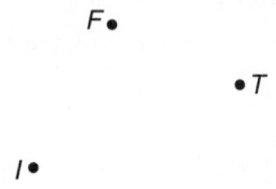

SRB 92

3. Draw and label line segment *GP*.

What is another name for $\overline{GP}$?

_____

SRB 90

4. Name as many rays as you can in the figure below.

Write their names.

_____

_____

SRB 91

5. Draw a quadrangle with 1 pair of parallel sides.

What kind of quadrangle is this?

_____

SRB 99 100

6. Put these numbers in order from least to greatest.

32,000 3,200

23,000 2,300

_____

_____

_____

_____

SRB 4

LESSON 1•8 Copying a Line Segment

Steps 1–4 below show you how to copy a line segment.

Step 1 You are given line segment *AB* to copy.

Step 2 Draw a line segment that is longer than line segment *AB*. Label one of its endpoints *C*.

Step 3 Open your compass so that the anchor is on one endpoint of line segment *AB* and the pencil point is on the other endpoint.

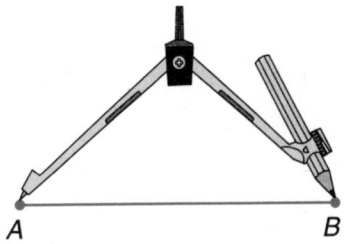

Step 4 *Without changing the compass opening,* place the anchor on point *C* on your second line segment. Make a mark that crosses this line segment. Label the point where the mark crosses the line segment with the letter *D*.

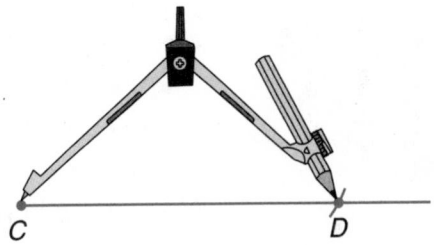

Line segment *CD* should be about the same length as line segment *AB*. Line segments *CD* and *AB* are **congruent.**

Use a compass and straightedge to copy the line segments shown below. For each problem, begin by drawing a line segment that is longer than the one given.

1.

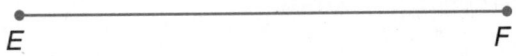

2.

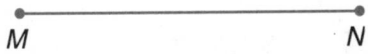

LESSON 1•8

Hexagons in Our World

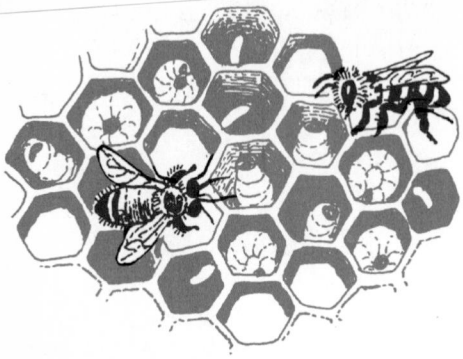

Hexagons are seen in the natural world and in things that people make and use. For example, bees make honeycombs with hexagonal shapes, and snowflakes suggest the shape of a hexagon.

Soccer balls are made up of regular hexagons and regular pentagons.

For many centuries, wonderful tile designs have been created all over the world, especially in Islamic art. As these pictures show, tile designs can be developed in many ways from a pattern that uses hexagons.

Many quilt and fabric designs come from dividing regular hexagons into triangles or rhombuses. You may have made designs like these with pattern blocks. Coloring a design often makes the design more interesting.

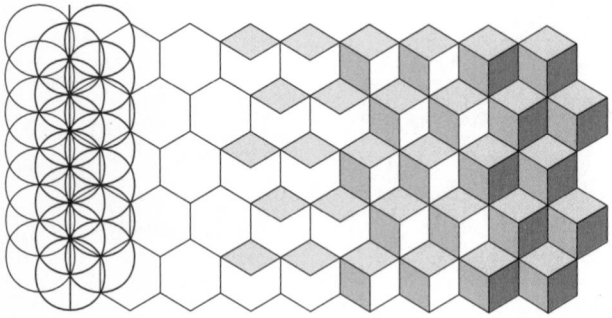

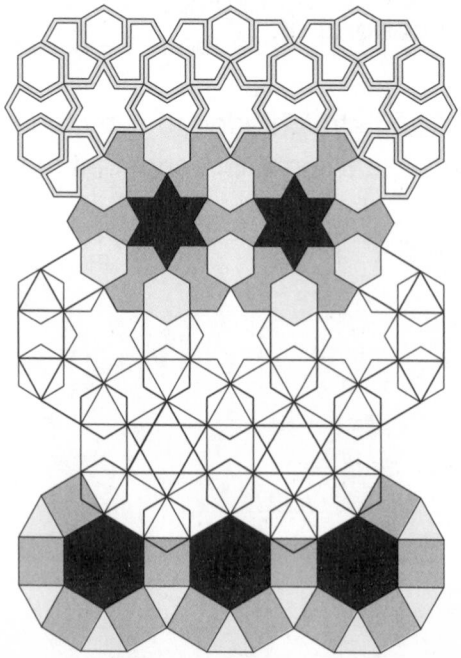

Date Time

LESSON 1•8

Constructing an Inscribed, Regular Hexagon

Follow each step below. Draw on a separate sheet of paper. Repeat these steps several times. Cut out your best work, and tape it onto the bottom of this page.

Step 1 Draw a circle. (Keep the same compass opening for Steps 2 and 3.) Draw a dot on the circle. Place the anchor of your compass on the dot and make a mark on the circle.

Step 2 Place the anchor of your compass on the mark you just made and make another mark on the circle.

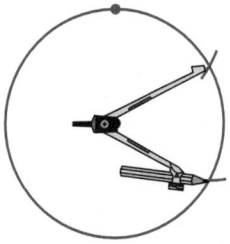

Step 3 Do this four more times to divide the circle into 6 equal parts. The 6th mark should be on the dot you started with or very close to it.

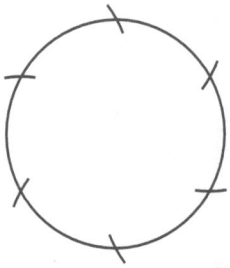

Step 4 With your straightedge, connect the 6 marks on the circle to form a regular hexagon. Use your compass to check that the sides of the hexagon are all about the same length.

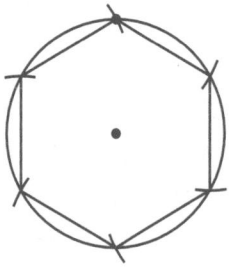

The hexagon is **inscribed** in the circle because each vertex of the hexagon is on the circle.

Date Time

LESSON 1•8

More Constructions

Construct a regular hexagon on a separate sheet of paper. Then divide the hexagon into 6 equilateral triangles. Use your compass to check that the sides of the 6 equilateral triangles are all about the same length.

Try this several times until you are satisfied with your work. Then cut out your best work and tape it in the space below.

Date Time

LESSON 1•8

Definition Match

Match each description of a geometric figure in Column I with its name in Column II. Some of the items in Column II do not have a match.

I	II
a. a polygon with 4 right angles and 4 sides of the same length	_____ octagon
	_____ rhombus
b. a polygon with 4 sides, none of which are the same length	_____ right angle
c. a quadrilateral with exactly 1 pair of opposite sides that is parallel	_____ trapezoid
	_____ hexagon
d. lines that never intersect	_____ square
e. a parallelogram with all sides the same length, but not a rectangle	_____ equilateral triangle
f. a polygon with 8 sides	_____ perpendicular lines
g. a polygon with 5 sides	_____ parallel lines
	_____ pentagon
h. an angle that measures 90°	_____ isosceles triangle
i. a triangle with all sides the same length	_____ quadrangle

Date Time

LESSON 1•8

Math Boxes

1. Subtract mentally.

   a. 14 − 9 = _____

   b. 13 − 8 = _____

   c. _____ = 18 − 9

   d. 17 − 8 = _____

   e. _____ = 11 − 7

   f. _____ = 15 − 6

2. Which of the shape(s) below are polygons? _____

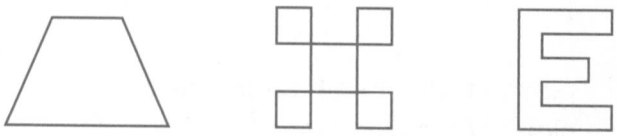

A B C

SRB 96

3. Draw a quadrangle that has 2 pairs of parallel sides and no right angles.

   What kind of quadrangle is this?

   _____

SRB 99 100

4. Circle the concave (nonconvex) polygon(s).

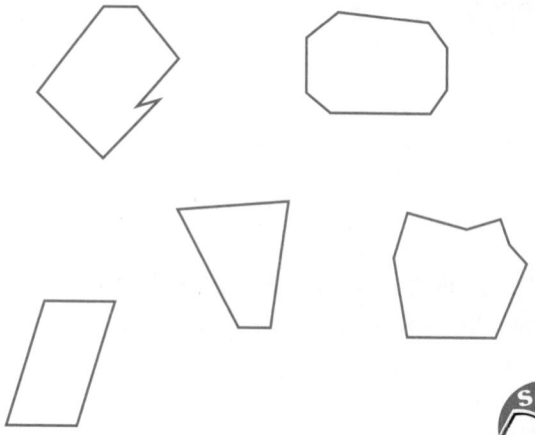

SRB 97

5. Draw and label ray *CA*.
   Draw point *R* on it.

   What is another name for ray *CA*?

   _____

SRB 91

6. In the numeral 30,516, what does the 3 stand for? Circle the best answer.

   A. 3,000

   B. 30

   C. 30,000

   D. 300,000

SRB 4

Date Time

LESSON 1•9

Math Boxes

1. Add.

a. 64
+ 32

b. 48
+ 96

SRB 10 11

2. Subtract.

a. 78
− 42

b. 81
− 36

SRB 12–15

3. Put these numbers in order from least to greatest.

46,000 40,600

4,600 4,006

_____

_____

_____

_____

SRB 4

4. In the numeral 78,965,

a. the 8 stands for _____.

b. the 6 stands for _____.

c. the 7 stands for _____.

d. the 9 stands for _____.

SRB 4

5. Use the following list of numbers to answer the questions:

12, 3, 15, 6, 12, 14, 6, 5, 9, 12

a. Which number is the least? _____

b. Which number is the greatest? _____

c. What is the difference between the least and greatest numbers? _____

d. Which number appears most often? _____

SRB 73

Date Time

LESSON 2·1

A Visit to Washington, D.C.

Refer to pages 267–269 in your *Student Reference Book*.

1. About how many people tour the White House every year? Check the best answer.

   ____ between 100 thousand and 1 million ____ between 1 million and 10 million

   ____ between 10 million and 100 million ____ between 100 million and 1 billion

2. About how many people ride the Washington Metrorail on an average weekday? Check the best answer.

   ____ between 100 thousand and 1 million ____ between 1 million and 10 million

   ____ between 10 million and 100 million ____ between 100 million and 1 billion

3. The Library of Congress adds about _____ items each day. About how many days does it take to add 50,000 items to the Library of Congress? _____

4. Write the year that each event happened. Then draw a dot for each event on the timeline below. Label the dot with the correct letter and date.

   A The year the Metrorail opened 1976

   B The year of the flight of the *Flyer* _____

   C The year the Washington Monument was completed _____

   D The year the Lincoln Memorial was dedicated _____

   E The year of the first nonstop flight across the Atlantic _____

   F The year the Jefferson Memorial was dedicated _____

   G The year of the first landing on the moon _____

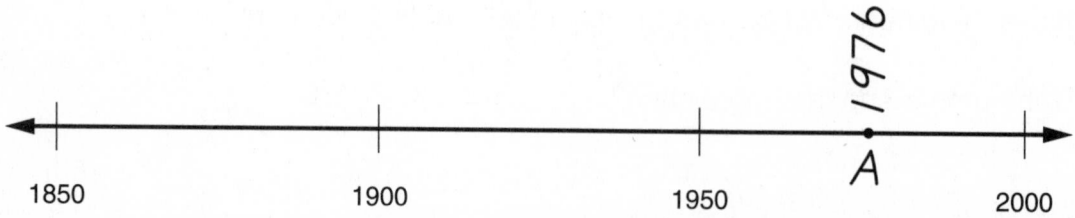

Date Time

LESSON 2·1

Math Boxes

1. Add mentally.

a. 3 + 5 = _____

b. 30 + 50 = _____

c. 300 + 500 = _____

d. _____ = 9 + 7

e. _____ = 90 + 70

f. _____ = 900 + 700

SRB 10 11

2. What is the value of the digit 5 in **5**60? 500

What is the value of the digit 7 in the numbers below?

a. 4**7**4 _____

b. **7**0,158 _____

c. 18**7**,943 _____

d. 2,**7**31,008 _____

SRB 4

3. Circle the pair of concentric circles.

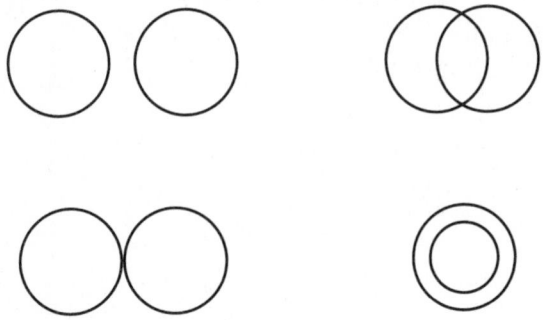

4. I am a 2-dimensional figure.
I have two pairs of parallel sides.
All my angles have the same measure.
All my sides are the same length.

What am I? _____

Use your Geometry Template to draw me.

SRB 100

5. A giant tortoise can live for about 150 years. An elephant can live for about 78 years. About how much longer can a giant tortoise live than an elephant? Fill in the circle next to the best answer.

(A) 88 years

(B) 238 years

(C) 72 years

(D) 228 years

SRB 178

6. Multiply mentally.

a. 2 × 1 = _____

b. _____ = 5 × 0

c. _____ = 5 × 2

d. 5 × 4 = _____

e. 3 × 10 = _____

SRB 16

Date Time

LESSON 2•2

Name-Collection Boxes

Write five names in each box below. Use as many different kinds of numbers (such as counting numbers, fractions, decimals, negative numbers) and different operations (+, −, ×, ÷) as you can. Draw a star next to the name you find most interesting.

SRB 149

1.

16
8×2
$970 - 954$
$10 + (60 \div 10)$

2.

24

3.

50

4.

100

Make up your own name-collection boxes. Use different kinds of numbers and operations.

5.

6.

Date Time

LESSON 2•2

Math Boxes

1. a. Write the largest number you can make with the digits 5, 2, 3, 0, 6, 0. Use each digit only once.

_____

b. Use the same digits and write the smallest number you can make. Do not start with 0.

_____

SRB 4

2. Add mentally or with a paper-and-pencil algorithm.

a. $\begin{array}{r} 37 \\ +\ 142 \\ \hline \end{array}$

b. $\begin{array}{r} 468 \\ +\ 394 \\ \hline \end{array}$

SRB 10 11

3. Draw a convex polygon.

SRB 97

4. Measure these line segments to the nearest centimeter.

a. _____

About _____ centimeters

b. _____

About _____ centimeters

SRB 128

5. Complete.

a. 4 ft = _____ in.

b. 4 ft = _____ yd _____ in.

c. 2 yd = _____ ft

d. 72 in. = _____ yd _____ ft

e. 6,756 in. = _____ ft

SRB 129

6. Divide mentally.

a. 9 ÷ 9 = _____

b. _____ = 12 ÷ 2

c. _____ = 20 ÷ 5

d. 40 ÷ 10 = _____

e. 14 ÷ 7 = _____

SRB 20

Date Time

LESSON 2•3

Place-Value Chart

SRB 4

Number	Hundred Millions	Ten Millions	Millions	Hundred Thousands	Ten Thousands	Thousands	Hundreds	Tens	Ones
	100M	10M	M	100K	10K	K	H	T	O

Date _____ Time _____

LESSON 2•3 Taking Apart, Putting Together

Complete.

1. In 574,

   5 is worth 500

   7 is worth _____

   4 is worth _____

2. In 9,027,

   9 is worth _____

   0 is worth _____

   2 is worth _____

3. In 280,743,

   8 is worth _____

   2 is worth _____

   4 is worth _____

4. In 56,010,837,

   6 is worth _____

   1 is worth _____

   5 is worth _____

5. In 705,622,463,

   5 is worth _____

   6 is worth _____

   7 is worth _____

6. In 123,456,789,

   4 is worth _____

   3 is worth _____

   2 is worth _____

Add.

7.
```
    900
     70
+     5
```

8.
```
 30,000
  7,000
     50
+     2
```

9.
```
 50,000,000
  9,000,000
     60,000
      2,000
        800
+        50
```

10.
```
 300,000,000
   9,000,000
     200,000
      70,000
          30
+          1
```

11. Think about why we need zeros when writing numbers. What would happen if you did not write the zero in the number 5,074?

_____

_____

Date Time

LESSON 2•3

Polygon Checklist

Place a check mark next to all of the statements that are true about each figure. Write an additional true statement for each figure.

SRB 93–100

1.

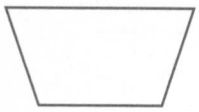

____ 1 pair of parallel sides

____ at least 1 right angle

____ quadrangle

____ polygon

____ concave

____ parallelogram

✓ ____ _____

2.

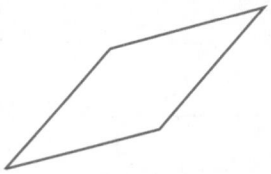

____ 4 sides of equal length

____ kite

____ square

____ parallelogram

____ convex

____ opposite sides parallel

✓ ____ _____

3.

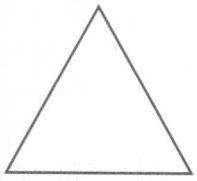

____ all sides of equal length

____ all angles of equal measure

____ one right angle

____ polygon

____ equilateral triangle

____ 1 pair of parallel sides

✓ ____ _____

4.

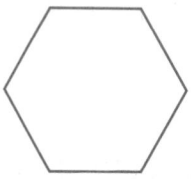

____ regular polygon

____ all sides of equal length

____ all angles of equal measure

____ pentagon

____ octagon

____ all angles smaller than right angles

✓ ____ _____

Date Time

LESSON 2·3

Math Boxes

1. Add mentally.

   a. 4 + 5 = _____

   b. 40 + 50 = _____

   c. 400 + 500 = _____

   d. _____ = 5 + 8

   e. _____ = 50 + 80

   f. _____ = 500 + 800

SRB 10 11

2. What is the value of the digit 8 in the numbers below?

   a. 5**8**4 _____

   b. 3**8**,067 _____

   c. 49,**8**41 _____

   d. **8**20,731 _____

   e. **8**,391,467 _____

SRB 4

3. Use your compass to draw a pair of concentric circles.

4. I am a 2-dimensional figure.
   I have two pairs of parallel sides.
   None of my angles is a right angle.
   All of my sides are the same length.

   What am I? _____

   Use your Geometry Template to draw me.

SRB 100

5. A sailfish can swim at a speed of 110 kilometers per hour. A tiger shark can swim at a speed of 53 kilometers per hour. How much faster can a sailfish swim than a tiger shark?

   _____ kilometers per hour

6. Multiply mentally.

   a. 8 × 1 = _____

   b. _____ = 9 × 0

   c. _____ = 5 × 6

   d. 5 × 5 = _____

   e. 7 × 10 = _____

SRB 16

Date Time

LESSON 2•4

Calculator "Change" Problems

1. Follow your teacher's directions to complete the "change" problems below. Use your calculator.

	Start with	Place of Digit	Change to	Operation	New Number
a.	570	Tens			
b.	409	Hundreds			
c.	54,463	Thousands			
d.	760,837	Tens			
e.	52,036,458	Ones			
f.		Ten Thousands			
g.		Millions			

2. Complete these calculator "change" problems on your own.

	Start with	Place of Digit	Change to	Operation	New Number
a.	893	Tens	3	–	
b.	5,489	Hundreds	7	+	
c.	94,732	Thousands	6	+	
d.	218,149	Ten Thousands	0	–	
e.	65,307,000	Millions	9	+	
f.	873,562,003	Ten Millions	1		
g.	103,070,651	Hundred Millions	8		

Date Time

LESSON 2•4

Math Boxes

1. What is the largest number you can make with the digits 3, 0, 3, 8, and 0? Fill in the circle next to the best answer.

   (A) 83,003

   (B) 83,030

   (C) 83,300

   (D) 80,033

SRB 4

2. Add mentally or with a paper-and-pencil algorithm.

   a. $\begin{array}{r} 145 \\ +\ 34 \\ \hline \end{array}$

   b. $\begin{array}{r} 297 \\ +136 \\ \hline \end{array}$

SRB 10 11

3. Draw a concave pentagon.

SRB 97

4. Measure these line segments to the nearest centimeter.

   a. _____

   About _____ centimeters

   b. _____

   About _____ centimeters

SRB 128

5. Complete.

   a. 14 in. = _____ ft _____ in.

   b. _____ in. = 2 ft

   c. _____ ft = 7 yd

   d. 1 yd 1 ft = _____ in.

   e. 413 ft = _____ yd _____ ft

SRB 129

6. Divide mentally.

   a. 16 ÷ 2 = _____

   b. 20 ÷ 10 = _____

   c. _____ = 40 ÷ 5

   d. 60 ÷ 10 = _____

   e. _____ = 45 ÷ 5

SRB 20

Date Time

LESSON 2•5

Counting Raisins

SRB 71–75

1. Use your $\frac{1}{2}$-ounce box of raisins. Complete each step when the teacher tells you. Stop after you complete each step.

   a. Don't open your box yet. **Guess** about how many raisins are in the box.

   About _____ raisins

   b. Open the box. Count the number of raisins in the top layer. Then **estimate** the total number of raisins in the box.

   About _____ raisins

   c. Now **count** the raisins in the box.

   How many? _____ raisins

2. Make a tally chart of the class data.

Number of Raisins	Number of Boxes

3. Find the following **landmarks** for the class data.

   a. What is the **maximum,** or largest, number of raisins found? _____

   b. What is the **minimum,** or smallest, number of raisins found? _____

   c. What is the **range?** (Subtract the minimum from the maximum.) _____

   d. What is the **mode,** or most frequent number of raisins found? _____

Try This

4. What is the **median** number of raisins found? _____

5. What is the **mean** number of raisins found? _____

Date Time

LESSON 2•5

Math Boxes

1. A number has

   6 in the hundreds place,
   1 in the millions place,
   2 in the tens place,
   8 in the hundred-thousands place,
   5 in the ones place,
   3 in the thousands place, and
   4 in the ten-thousands place.

   Write the number.

   ___, ___ ___ ___, ___ ___ ___

SRB 4

2. Write five names for 34.

34

SRB 149

3. Write >, <, or = to make each number sentence true.

   a. 14 _____ 26

   b. 3,003 _____ 3,300

   c. 12 + 12 _____ 24

   d. 200 − 50 _____ 100

   e. 30 + 30 _____ 50 + 10

SRB 148 149

4. Name the two pairs of parallel sides in parallelogram *HIJK*.

   _____ and _____

   _____ and _____

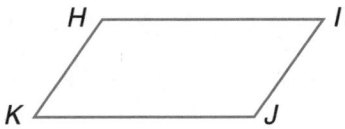

SRB 94

5. Measure these line segments to the nearest $\frac{1}{2}$ centimeter.

   a. _____

   About _____ centimeters

   b. _____

   About _____ centimeters

SRB 128

6. Multiply mentally.

   a. 5 × 7 = _____

   b. 3 × _____ = 18

   c. _____ × 7 = 56

   d. 9 × _____ = 45

   e. 8 × 4 = _____

SRB 16

Date Time

LESSON 2•6

Family Size

Follow your teacher's directions and complete each step.

1. How many people are in your family? _____ people
   Write the number on a stick-on note.

2. Make a line plot of the family-size data for the class.
   Use **X**s in place of stick-on notes.

Class Data on Family Size

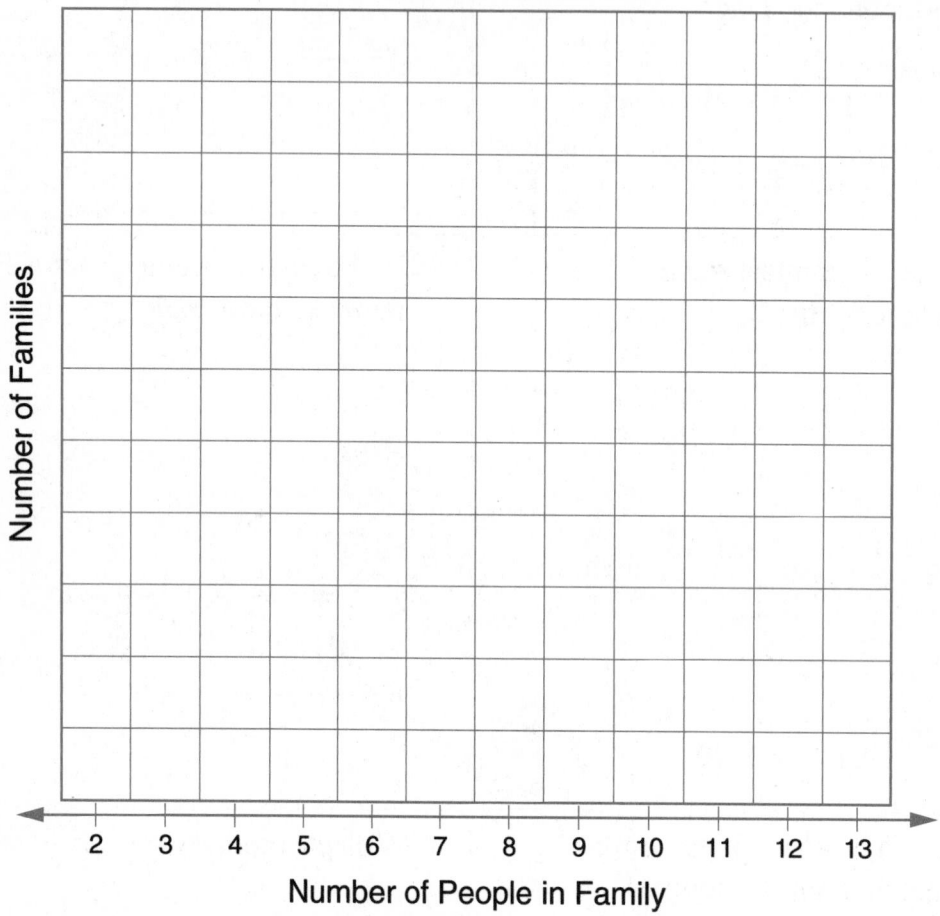

3. Find the following landmarks for the class data:

   a. What is the **maximum** (largest) number of people in a family? _____ people

   b. What is the **minimum** (smallest) number of people in a family? _____ people

   c. What is the **range?** (Subtract the minimum from the maximum.) _____ people

   d. What is the **mode** (most frequent family size)? _____ people

4. What is the **median** family size for the class? _____ people

Date Time

LESSON 2•6

Math Boxes

1. Add mentally.

   a. 2 + 7 = _____

   b. 20 + 70 = _____

   c. 200 + 700 = _____

   d. _____ = 8 + 4

   e. _____ = 80 + 40

   f. _____ = 800 + 400

SRB 10 11

2. Find the median of the data set.

   3, 45, 13, 15, 3, 7, 19

   Fill in the circle next to the best answer.

   (A) 3

   (B) 13

   (C) 15

   (D) 45

SRB 73

3. Subtract mentally or with a paper-and-pencil algorithm.

   a. 147 − 56

   b. 531 − 246

SRB 12–15

4. Write 4,007,392 in words.

   _____

   _____

   _____

   _____

   _____

SRB 4

5. A royal python can be 35 feet long. An anaconda can be 28 feet long. What would be their combined length, end-to-end?

   _____ feet

6. Tell whether each number sentence is true or false.

   a. 14 + 7 = 22 _____

   b. 36 = 15 + 5 _____

   c. 45 − 12 = 33 _____

   d. 27 = 40 − 13 _____

SRB 148

Date Time

LESSON 2•7

Partial-Sums Addition

Write a number model for a ballpark estimate. Solve Problems 1–3 using the partial-sums method. Solve Problems 4–6 using any method. Compare your answer with your estimate to see if your answer makes sense.

SRB 10

1. 76 + 38 Ballpark estimate: _____	**2.** 647 + 936 Ballpark estimate: _____	**3.** 1,672 + 3,221 Ballpark estimate: _____
4. 66 + 28 Ballpark estimate: _____	**5.** 736 + 645 Ballpark estimate: _____	**6.** 7,854 + 4,550 Ballpark estimate: _____

Try This

7. Name three 4-digit numbers whose sum is 17,491.

_____ + _____ + _____ = 17,491

Date Time

LESSON 2•7

Column Addition

Write a number model for a ballpark estimate. Solve Problems 1–3 using the column-addition method. Solve Problems 4–6 using any method. Compare your answer with your estimate to see if your answer makes sense.

SRB 11

1. 94 + 47 Ballpark estimate: _____	**2.** 385 + 726 Ballpark estimate: _____	**3.** 2,538 + 4,179 Ballpark estimate: _____
4. 49 + 33 Ballpark estimate: _____	**5.** 469 + 946 Ballpark estimate: _____	**6.** 4,614 + 6,058 Ballpark estimate: _____

Try This

7. Name four 4-digit numbers whose sum is 15,706.

_____ + _____ + _____ + _____ = 15,706

Date Time

LESSON 2•7

Math Boxes

1. A number has
   3 in the millions place,
   1 in the ones place,
   8 in the thousands place,
   9 in the ten-thousands place,
   0 in the tens place,
   6 in the hundred-thousands place, and
   5 in the hundreds place.

   Write the number.

   ___, ___ ___ ___, ___ ___ ___

SRB 4

2. Write five names for 100.

100

SRB 149

3. Write >, <, or = to make each number sentence true.

   a. 16 + 11 _____ 47

   b. 206 _____ 602

   c. 150 − 50 _____100

   d. 62 + 10 + 10 _____ 62 − 10 − 10

   e. 423,726 _____ 413,999

SRB 148 149

4. Draw a parallelogram. Label the vertices so that side *AB* is parallel to side *CD*.

SRB 99 100

5. Measure these line segments to the nearest $\frac{1}{2}$ centimeter.

   a. _____

   About _____ cm

   b. _____

   About _____ cm

SRB 128

6. Multiply mentally.

   a. 5 × 8 = _____

   b. 2 × _____ = 16

   c. 7 × _____ = 21

   d. _____ × 9 = 54

   e. 8 × 3 = _____

SRB 16

Date Time

LESSON 2•8

Math Boxes

1. Add mentally.

   a. 2 + 4 = _____

   b. 20 + 40 = _____

   c. 200 + 400 = _____

   d. _____ = 8 + 6

   e. _____ = 80 + 60

   f. _____ = 800 + 600

SRB 10 11

2. Find the following landmarks for this set of numbers: 12, 16, 23, 15, 16, 19, 18.

   a. median _____

   b. mode _____

   c. maximum _____

   d. minimum _____

   e. range _____

   f. mean _____

SRB 73–75

3. Subtract mentally or with a paper-and-pencil algorithm.

   a. 231 − 84

   b. 603 − 466

SRB 12–15

4. Write 8,042,176 in words.

   _____

   _____

   _____

   _____

   _____

SRB 4

5. An ostrich can weigh about 345 pounds. An emu can weigh about 88 pounds. How much would they weigh together?

   _____ pounds

6. Tell whether each number sentence is true or false.

   a. 18 + 9 = 37 _____

   b. 29 = 17 + 12 _____

   c. 42 − 15 = 27 _____

   d. 17 = 40 − 24 _____

   e. 154 − 65 = 99 _____

SRB 148

Date Time

LESSON 2·8

Head Sizes

Ms. Woods owns a clothing store. She is trying to decide how many children's hats to stock in each possible size. Should she stock the same number of hats in each size? Or should she stock more hats in some sizes and fewer in others?

SRB 73

Help Ms. Woods decide. Pretend that she has asked each class in your school to collect and organize data about students' head sizes. She plans to combine the data and then use it to figure out how many hats of each size to stock.

As a class, collect and organize data about one another's head sizes.

1. Ask your partner to help you measure the distance around your head.
   - Wrap the tape measure once around your head.
   - See where the tape touches the end tip of the tape measure.
   - Read the mark where the tape touches the end tip.
   - Read this length to the nearest $\frac{1}{2}$ centimeter.

Record your head size. About _____ cm

2. What is the median head size for the class? About _____ cm

3. Find the following landmarks for the head-size data shown in the bar graph on journal page 47.

   Minimum: _____ Maximum: _____ Range: _____

   Mode: _____ _____ Median: _____

4. How would the landmarks above help Ms. Woods, a clothing store owner, decide how many baseball caps of each size to stock?

_____

_____

_____

_____

_____

Date Time

LESSON 2•8

Head Sizes *continued*

Make a bar graph of the head-size data for the class.

SRB 76

title

label

label

Date Time

LESSON 2•8

How to Construct a Kite

Follow the steps to construct a kite in the space below.

SRB 100

Step 1 Draw points *A* and *B*. (*See example below.*)

Step 2 Set your compass opening so that it is a little more than half the distance between *A* and *B*. Place the point of the compass on *A* and draw an arc. *Without changing the compass opening,* place the point of the compass on *B* and draw a second arc that intersects the first arc. Label the point where the two arcs meet *C*.

Step 3 Change your compass opening. Set it so that it is almost the full distance between *A* and *B* (about $\frac{4}{5}$ of the width). Repeat Step 2 as shown in the picture, and label the new point of intersection *D*.

Step 4 With your straightedge, connect the 4 points to form a quadrangle.

Example:

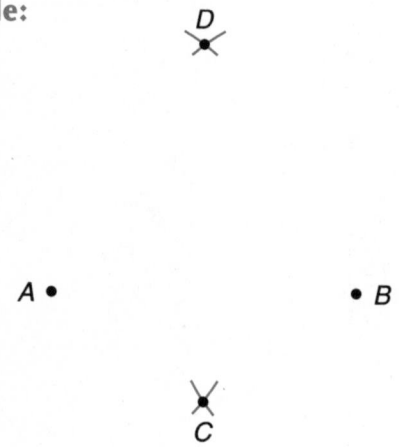

Steps 1–3

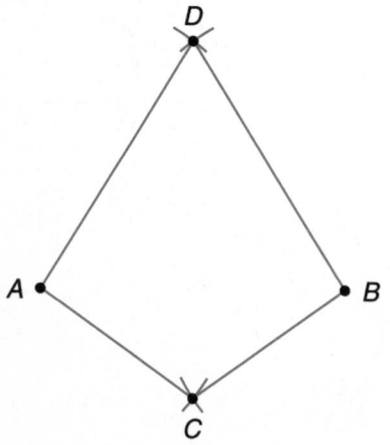

Step 4

Draw your kite here.

Date Time

LESSON 2•9

Trade-First Subtraction

Write a number model for a ballpark estimate. Solve Problems 1–3 using the trade-first method. Solve Problems 4–6 using any method. Use your ballpark estimates to see if your answers make sense.

SRB 12 13

1. 58 − 39 Ballpark estimate: _____	**2.** 600 − 379 Ballpark estimate: _____	**3.** 2,936 − 1,657 Ballpark estimate: _____
4. 73 − 37 Ballpark estimate: _____	**5.** 900 − 461 Ballpark estimate: _____	**6.** 2,468 − 1,789 Ballpark estimate: _____

Try This

7. Name two 3-digit numbers whose difference is 357.

_____ − _____ = 357

Date Time

LESSON 2•9

Partial-Differences Subtraction

Write a number model for a ballpark estimate. Solve Problems 1–3 using the partial-differences method. Solve Problems 4–6 using any method. Use your ballpark estimates to see if your answers make sense.

1. 86 − 37 Ballpark estimate: _____	**2.** 900 − 485 Ballpark estimate: _____	**3.** 7,584 − 2,806 Ballpark estimate: _____
4. 61 − 26 Ballpark estimate: _____	**5.** 400 − 271 Ballpark estimate: _____	**6.** 3,681 − 1,803 Ballpark estimate: _____

Try This

7. Name two 4-digit numbers whose difference is 4,203.

_____ − _____ = 4,203

LESSON 2•9

Math Boxes

1. A number has

   6 in the tens place,
   9 in the millions place,
   4 in the thousands place,
   3 in the ten-thousands place,
   2 in the hundred-thousands place,
   0 in the ones place, and
   8 in the hundreds place.

   Write the number.

   ___, ___ ___ ___, ___ ___ ___

SRB 4

2. Write five names for 1,000.

1,000

SRB 149

3. Which number sentence is true? Fill in the circle next to the best answer.

   (A) $34 - 4 = 31$

   (B) $812 < 218$

   (C) $423 + 20 > 443$

   (D) $123 + 5 + 5 = 113 + 20$

SRB 148 149

4. Draw a polygon that has no parallel sides.

SRB 94

5. Measure these line segments to the nearest $\frac{1}{2}$ centimeter.

   a. _____

   About _____ cm

   b. _____

   About _____ cm

SRB 128

6. Multiply mentally.

   a. _____ $= 5 \times 4$

   b. $5 \times 6 =$ _____

   c. $9 \times 3 =$ _____

   d. $4 \times$ _____ $= 24$

   e. $6 \times$ _____ $= 48$

SRB 16

Date Time

LESSON 2•10

Math Boxes

1. Divide mentally.

   a. 35 ÷ 5 = _____

   b. 56 ÷ _____ = 8

   c. 32 ÷ _____ = 4

   d. 24 ÷ _____ = 6

   e. 72 ÷ 8 = _____

   f. 40 ÷ 5 = _____

SRB 20

2. Multiply mentally.

   a. 4 × 7 = _____

   b. 3 × _____ = 15

   c. 7 × _____ = 42

   d. 9 × _____ = 36

   e. 6 × 0 = _____

   f. 1 × 9 = _____

SRB 16

3. Complete the square facts.

   a. 64 ÷ 8 = _____

   b. 49 ÷ 7 = _____

   c. _____ = 4 × 4

   d. _____ = 3 × 3

   e. 25 ÷ 5 = _____

SRB 158

4. Tell whether each number sentence is true or false.

   a. 46 + 12 = 53 _____

   b. 36 = 22 + 14 _____

   c. 13 = 84 − 71 _____

   d. 52 − 20 = 34 _____

SRB 148

5. A grizzly bear can weigh 786 pounds. An American black bear can weigh 227 pounds. What is their combined weight?

   _____ pounds

6. On average, India produces about 855 movies per year. The United States produces about 762 movies. On average, how many fewer movies per year does the United States produce than India?

   _____

Date Time

LESSON 3•1

"What's My Rule?"

SRB 162–166

Complete the "What's My Rule?" tables and state the rules.

1.

in	out
30	
80	
20	
150	
290	

2.

in	out
	50
	210
	20
	270
	340

3.

in	out
49	72
151	
272	295
	611
	503

4. Rule: There are 12 inches in 1 foot.

in	out
3	36
	60
10	
	264
	720

Try This

5. Rule: _____

in	out
17	−8
12	
27	2
−5	
	0

6. Create your own.

Rule: _____

in	out

Date Time

LESSON 3•1

A Polygon Alphabet

Try reading this message:

ALL OF THESE LETTERS ARE POLYGONS.

1. Use a straightedge to design a polygon letter for each of the letters shown below. You'll have to simplify, because a polygon can't have any curves, and it can't have any "holes."

   For example, if you look at the letter "P," you see that there is no opening in the upper part. Making it look like this, , would make it easier to read, but it would not be a polygon.

B	C	D
F	M	X

2. Which of the letters you drew are nonconvex (concave) polygons? _____
   How do you know?

   _____

3. Do any of the letters you drew have special names as polygons? Explain.

   _____

   _____

Try This

4. On a separate sheet of paper, design polygon letters for the rest of the uppercase (capital) letters in the alphabet, the 26 lowercase (small) letters, or the 10 digits (0–9).

Date Time

LESSON 3•1

Math Boxes

1. Write >, <, or = to make each number sentence true.

   a. 1 million _____ 100,000

   b. 73,099 _____ 71,999

   c. 304,608 _____ 304,809

   d. 5,682 _____ 7 hundred

   e. 5,000,236 _____ 5,000,099

SRB 6 149

2. Number of spelling words correct for 10 students on the spelling test:

   25, 19, 16, 25, 18, 19, 25, 24, 25, 23

   a. What is the range for this set of numbers? _____

   b. What is the median? _____

SRB 73

3. Make a ballpark estimate. Write a number model to show your strategy.

   a. 3,389 + 2,712

   _____ + _____ = _____

   b. 3,452 − 1,147

   _____ − _____ = _____

SRB 181

4. Complete.

   a. 21 ft = _____ yd

   b. 4 ft = _____ in.

   c. 16 ft = _____ yd _____ ft

   d. 2 yd 2 ft = _____ in.

   e. _____ ft _____ in. = 568 in.

SRB 129

5. Complete.

   a. 7, 15, 23, ____, ____, ____

   Rule: _____

   b. 49, 42, ____, 28, ____, ____

   Rule: _____

   c. ____, ____, 53, 59, ____, 71

   Rule: _____

SRB 160 161

6. Solve mentally or with a paper-and-pencil algorithm.

   a. $3.56
      + $2.49

   b. $6.25
      − $5.01

SRB 34–37

Date Time

LESSON 3•2

Multiplication/Division Facts Table

SRB 16 20

*, /	1	2	3	4	5	6	7	8	9	10
1	1	2	3	4	5	6	7	8	9	10
2	2	4	6	8	10	12	14	16	18	20
3	3	6	9	12	15	18	21	24	27	30
4	4	8	12	16	20	24	28	32	36	40
5	5	10	15	20	25	30	35	40	45	50
6	6	12	18	24	30	36	42	48	54	60
7	7	14	21	28	35	42	49	56	63	70
8	8	16	24	32	40	48	56	64	72	80
9	9	18	27	36	45	54	63	72	81	90
10	10	20	30	40	50	60	70	80	90	100

Date Time

LESSON 3•2

Math Boxes

1. The numbers 28, 35, and 42 are all multiples of __. Circle the best answer.

   A 7

   B 4

   C 6

   D 2

2. Complete the "What's My Rule?" table and state the rule.

   Rule: _____

in	out
236	331
682	777
	486
938	
647	

3. Earth is covered by a rocky outer layer called the *crust,* which is made up of many elements.

   a. Is there more aluminum or silicon in Earth's crust?

   _____

   b. What percentage of Earth's crust is aluminum?

   _____

   c. Which element makes up most of Earth's crust?

   _____

Elements Found in Earth's Crust (percent by weight)

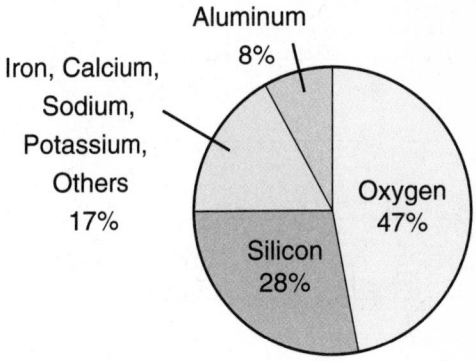

4. Name as many line segments as you can in the figure below.

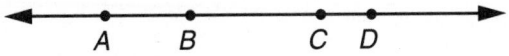

_____

5. Put these numbers in order from smallest to largest.

   0.6 0.06 0.43 0.9

   _____

   _____

   _____

   _____

Date Time

LESSON 3•3

Patterns in Multiplication Facts

Math Message

Look at the table on journal page 56.

1. Find a pattern in the 9s multiplication facts. Describe the pattern.

2. Find a pattern in the 5s multiplication facts. Describe the pattern.

3. What other patterns can you find in the multiplication facts? Write about some of them.

Date Time

LESSON 3•3

Math Boxes

1. Write >, <, or = to make each number sentence true.

   a. 45,699 _____ 45,609

   b. 67,749 _____ 66,749

   c. 208,775 _____ 200 million

   d. 1,000,000 _____ 858,192

   e. 2 million _____ 20,000,000

SRB 6 149

2. Number of days it took 10 students to complete their science projects:

   6, 4, 10, 11, 8, 6, 14, 9, 3, 12

   a. What is the range for this set of numbers?

   _____

   b. What is the median?

   _____

SRB 73

3. Make a ballpark estimate. Write a number model to show your strategy.

   a. 1,459 + 291

   _____ + _____ = _____

   b. 1,381 − 646

   _____ − _____ = _____

SRB 181

4. Complete.

   a. 3 yd = _____ ft

   b. 4 ft = _____ in.

   c. 54 in. = _____ ft _____ in.

   d. $\frac{1}{2}$ yd = _____ ft _____ in.

   e. $17\frac{1}{2}$ yd = _____ in.

SRB 129

5. Complete.

   a. 20, 35, 50, _____, _____, _____

   Rule: _____

   b. _____, 68, _____, 94, _____, 120

   Rule: _____

   c. 58, _____, _____, _____, _____, −2

   Rule: _____

SRB 160 161

6. Solve mentally or with a paper-and-pencil algorithm.

   a. $\begin{array}{r} \$10.97 \\ +\ \$15.60 \\ \hline \end{array}$

   b. $\begin{array}{r} \$4.56 \\ -\ \$2.07 \\ \hline \end{array}$

SRB 34–37

Date Time

LESSON 3•4

Math Boxes

1. Complete.

a. Name four multiples of 5.

_____, _____, _____, _____

b. Name four multiples of 9.

_____, _____, _____, _____

SRB 9

2. Complete the "What's My Rule?" table and state the rule.

Rule: _____

in	out
502	346
1,238	1,082
	927
871	
1,600	

SRB 162–166

3. Mr. Rosario's fourth graders collected data about their favorite fruits.

a. Which kind of fruit do most students prefer? _____

b. Which kind is preferred least? _____

c. Do more people prefer apples or peaches? _____

d. About $\frac{1}{4}$ of the students prefer _____.

Favorite Fruits

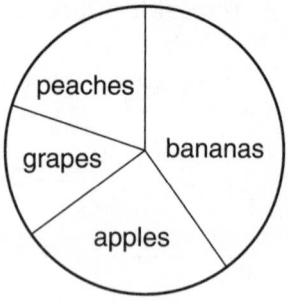

4. Name as many line segments as you can in the figure below.

_____

SRB 90

5. Put these numbers in order from smallest to largest.

0.7 0.007 0.5 0.63

_____

_____

_____

_____

SRB 32 33

Date Time

LESSON 3•5

Multiplication and Division

SRB 16 20

Equivalents			
$3 * 4$	$12 / 3$	$12 \div 3$	$3 < 5$ (< means "is less than")
3×4	$\frac{12}{3}$	$3\overline{)12}$	$5 > 3$ (> means "is greater than")

1. Choose 3 Fact Triangles. Write the fact family for each.

_____ * _____ = _____	_____ × _____ = _____	_____ * _____ = _____
_____ * _____ = _____	_____ × _____ = _____	_____ * _____ = _____
_____ / _____ = _____	_____ / _____ = _____	_____ ÷ _____ = _____
_____ / _____ = _____	_____ / _____ = _____	_____ ÷ _____ = _____

2. Solve each division fact.

a. 27 / 3 = _____ Think: How many 3s in 27?

b. _____ = 45 / 5 Think: How many 5s in 45?

c. 36 ÷ 6 = _____ Think: 6 times what number equals 36?

d. 24 / 8 = _____ Think: 8 times what number equals 24?

Try This

3. A cashier has 5 rolls of quarters and 6 rolls of dimes in the cash register. Each roll of quarters is worth $10, and each roll of dimes is worth $5.

a. How much are the rolls of quarters and dimes worth in all? $_____

b. How many quarters are in 1 roll? _____ quarters

c. How many quarters are in the 5 rolls? _____ quarters

d. How many dimes are in 1 roll? _____ dimes

e. How many dimes are in the 6 rolls? _____ dimes

f. There is also $7.50 worth of half-dollars in the cash register. How many half-dollars is that? _____ half-dollars

Date Time

LESSON 3•5

Math Boxes

1. Write >, <, or = to make each number sentence true.

   a. 5,389 _____ 3,389

   b. 70,642 _____ 70,699

   c. 6 million _____ 6,000,000

   d. 8,000,032 _____ 8 million, 32 thousand

   e. 400 + 30 + 5 _____ 4,000 + 30 + 5

SRB 6 149

2. The number of glasses of milk drunk by 10 students in a week:

   16, 13, 15, 20, 8, 10, 15, 12, 10, 18

   What is the range? Circle the best answer.

   A 8

   B 20

   C 12

   D 14

SRB 73

3. Make a ballpark estimate. Write a number model to show your strategy.

   a. 13,685 − 8,379

   _____ − _____ = _____

   b. 7,602 − 3,213

   _____ − _____ = _____

SRB 181

4. Complete.

   a. 31 in. = _____ ft _____ in.

   b. 17 ft = _____ yd _____ ft

   c. _____ ft = 14 yd

   d. _____ in. = 2 yd 1 ft

   e. $2\frac{1}{4}$ miles = _____ ft

SRB 129

5. Complete.

   a. 7, 11, 15, _____, _____, _____

   Rule: _____

   b. _____, _____, _____, 22, 25, 28

   Rule: _____

   c. _____, 14, _____, 28, _____, 42

   Rule: _____

SRB 160 161

6. Solve mentally or with a paper-and-pencil algorithm.

   a. $\begin{array}{r} \$2.27 \\ +\ \$4.96 \\ \hline \end{array}$

   b. $\begin{array}{r} \$5.00 \\ -\ \$3.64 \\ \hline \end{array}$

SRB 34–37

Date Time

LESSON 3•6

Math Boxes

1. Complete the name-collection box.

125

SRB 149

2. A number has

2 in the tens place,
8 in the hundred-thousands place,
5 in the millions place,
7 in the hundreds place,
9 in the ones place,
4 in the thousands place, and
1 in the ten-thousands place.

Write the number:

____, ____ ____ ____, ____ ____ ____

SRB 4

3. a. Measure line segment *PQ* to the nearest inch.

About _____ inches

b. Measure line segment *RS* to the nearest $\frac{1}{2}$ inch.

About _____ inches

SRB 128

4. Solve mentally or with a paper-and-pencil algorithm.

a.
$$\begin{array}{r} 729 \\ +\ 432 \\ \hline \end{array}$$

b.
$$\begin{array}{r} 9{,}004 \\ -\ \ 515 \\ \hline \end{array}$$

SRB 10–15

5. Riley estimated the height of his classroom ceiling. Circle the best estimate.

A 7 m

B 3 m

C 20 m

D 15 m

SRB 130

Date Time

LESSON 3•7

Measuring Air Distances

1. Estimate which city listed below is the closest to Washington, D.C. _____

2. Estimate which city is the farthest. _____

3. Measure the shortest distance between Washington, D.C., and each of the cities shown in the table below. Use the globe scale to convert these measurements to approximate air distances.

   a. Record the globe scale. _____ inch → _____ miles

   b. Complete the table.

Distance from Washington, D.C., to	Measurement on Globe (to the nearest $\frac{1}{2}$ inch)	Air Distance (estimated number of miles)
Cairo, Egypt		
Mexico City, Mexico		
Stockholm, Sweden		
Moscow, Russia		
Tokyo, Japan		
Shanghai, China		
Sydney, Australia		
Warsaw, Poland		
Cape Town, South Africa		
Rio de Janeiro, Brazil		
Choose a city. _____		

4. Explain how you used the globe scale to estimate the air distance between Washington, D.C., and Mexico City, Mexico.

_____

_____

_____

_____

Date Time

LESSON 3•7

Math Boxes

1. If 1 centimeter on a map represents 20 kilometers, then

   a. 2 cm represent _____ km.

   b. 5 cm represent _____ km.

   c. 8 cm represent _____ km.

   d. 3.5 cm represent _____ km.

   e. 6.5 cm represent _____ km.

SRB 145

2. Complete the "What's My Rule?" table and state the rule.

   Rule: _____

in	out
40	5
24	3
64	
	4
	9

SRB 162–166

3. A rock collector has 136 rocks in her collection. She took them to a geologist who said that 57 of them are volcanic. How many of them are not volcanic?

   _____

4. Solve the riddle. Then use your Geometry Template to draw the shape.

   I am a four-sided polygon.

   My two short sides are the same length.

   My two long sides are the same length.

   The sides of the same length are next to each other.

   What am I? _____

SRB 100

5. Make a ballpark estimate. Write a number model to show your strategy.

   $2.83 + $0.92 + $3.07 + $7.91

   _____

   _____

SRB 181

6. Complete.

   a. 0.1, 0.2, 0.3, _____, _____, _____

      Rule: _____

   b. 1.4, 1.6, 1.8, _____, _____, _____

      Rule: _____

   c. 3, 2.5, 2, _____, _____, _____

      Rule: _____

SRB 160 161

Date Time

LESSON 3•8

Flying to Cairo

SRB 175–178

Pretend that you are flying from Washington, D.C., to Cairo, Egypt. You have a choice of flying by way of Amsterdam, in the Netherlands, or by way of Rome, Italy. The air distances are shown in the table.

Air Distance between Cities (in miles)

	Amsterdam	Rome
Washington, D.C.	3,851	4,497
Cairo	2,035	1,326

If you fly first class, your ticket will cost $9,250.
If you fly economy class, you will save $6,500.

1. About how many more miles is it from Washington, D.C., to Rome than from Washington, D.C., to Amsterdam?

_____
(number model)

Answer: About _____ miles

2. What is the total distance from Washington, D.C., to Cairo by way of Amsterdam?

_____
(number model)

Answer: About _____ miles

3. What is the total distance from Washington, D.C., to Cairo by way of Rome?

_____
(number model)

Answer: About _____ miles

LESSON 3•8

Flying to Cairo *continued*

4. About how many fewer miles will you fly if you go by way of Rome rather than Amsterdam?

_____
(number model)

Answer: About _____ miles

Try This

5. A group of six friends spent $29,500 to fly round-trip from Washington, D.C., to Cairo.

   a. How many members of the group flew first class and how many flew economy?

   Answer: _____ first class and _____ economy

   b. Explain what you did to get your answer.

   _____

   _____

   _____

   _____

   _____

6. One of the flights to Amsterdam leaves Washington, D.C., at 1:15 P.M. It lands in Amsterdam 8 hours and 45 minutes later.

   a. At what time does it land, Washington, D.C., time?

   Answer: _____ (A.M. or P.M.?)

   b. At what time does it land, Amsterdam time? (Use the time zones map on pages 276 and 277 in the World Tour section of your *Student Reference Book*.)

   Answer: About _____ (A.M. or P.M.?)

Date Time

LESSON 3•8

Math Boxes

1. Complete the name-collection box.

480

SRB 149

2. A number has

1 in the tens place,
3 in the hundreds place,
5 in the ones place,
7 in the hundred-thousands place,
9 in the thousands place,
2 in the ten-thousands place, and
4 in the millions place.

Write the number:

___, ___ ___ ___, ___ ___ ___

SRB 4

3. a. Measure the line segment to the nearest inch.

About _____ in.

b. Measure the line segment to the nearest $\frac{1}{2}$ inch.

About _____ in.

SRB 128

4. Solve with a paper-and-pencil algorithm.

a. $\begin{array}{r} 604 \\ +\ 817 \\ \hline \end{array}$

b. $\begin{array}{r} 3{,}005 \\ -\ 686 \\ \hline \end{array}$

SRB 10–15

5. Sara estimated the length of her arm. Circle the best estimate.

A 5 centimeters

B 50 centimeters

C 100 centimeters

D 150 centimeters

SRB 130

Date Time

LESSON 3•9

Number Sentences

SRB 148

Tell whether each number sentence below is true or false. Write T for true or F for false. If it is not possible to tell, write ? on the answer blank.

1. $7 < 3 + 1$ _____
2. $6 = 36 \div 6$ _____
3. $80 - ? = 40$ _____
4. $28 - 16 = 12$ _____
5. $0 = 4 / 4$ _____
6. $2 * 7$ _____
7. $14 \times 3 < 19 \times 2$ _____
8. $144 + 76 = 880 \div 4$ _____
9. Make up two true number sentences and two false number sentences.
   a. true _____
   b. true _____
   c. false _____
   d. false _____
10. Make up three true number sentences and three false number sentences. Mix them up. Ask your partner to write whether each sentence is true or false.

Example: $4 * 7 = 34 - 6$ T

a. _____ _____

b. _____ _____

c. _____ _____

d. _____ _____

e. _____ _____

f. _____ _____

Date Time

LESSON 3•9

Math Boxes

1. If 1 inch on a map represents 30 miles, what would 3 inches represent? Circle the best answer.

   A 10 miles

   B 60 miles

   C 90 miles

   D 300 miles

SRB 145

2. Complete the "What's My Rule?" table and state the rule.

   Rule: _____

in	out
45	5
81	9
	3
	4
72	

SRB 162–166

3. The Statue of Chief Crazy Horse in South Dakota is 563 feet tall. The Statue of Liberty is 151 feet tall. What is the difference in height of the two statues?

   _____ feet

4. Solve the riddle. Then use your Geometry Template to trace the shape.

   I am a polygon.

   All my angles have the same measure.

   Each of my 5 sides has the same measure.

   What am I?

   _____

SRB 97

5. Make a ballpark estimate. Write a number model to show your strategy.

   $2.50 + $0.75 + $3.85 + $12.70

   _____

   _____

SRB 181

6. Complete.

   a. 5.05, 5.06, 5.07, _____, _____, _____

      Rule: _____

   b. 4, 3.8, 3.6, _____, _____, _____

      Rule: _____

   c. 2.7, 3.2, 3.7, _____, _____, _____

      Rule: _____

SRB 160 161

Date Time

LESSON 3•10

Parentheses in Number Sentences

Part 1

Make a true sentence by filling in the missing number.

1. a. (30 − 15) ∗ 2 = _____ b. 30 − (15 ∗ 2) = _____

2. a. _____ = 28 / (14 / 2) b. _____ = (28 / 14) / 2

3. a. (6 + 8) / (2 − 1) = _____ b. 6 + (8 / 2) − 1 = _____

Part 2

Make a true sentence by inserting parentheses.

4. a. 4 × 9 − 2 = 34 b. 4 × 9 − 2 = 28

5. a. 24 = 53 − 11 + 18 b. 60 = 53 − 11 + 18

6. a. 12 / 4 + 2 = 2 b. 12 / 4 + 2 = 5

7. a. 55 = 15 + 10 × 4 b. 100 = 15 + 10 × 4

Try This

Make a true sentence by inserting two sets of parentheses in each problem.

8. a. 10 − 4 / 2 ∗ 3 = 24 b. 10 − 4 / 2 ∗ 3 = 1

Part 3

Pretend you are playing a game of *Name That Number* with only 3 cards per hand. To name the target number, use all 3 numbers and any operations you want. For each problem, write a true number sentence containing parentheses using the 3 numbers and the target number.

9. Use: 2, 5, 15 Target number: 5 _____

10. Use: 3, 4, 5 Target number: 17 _____

11. Use: 1, 3, 11 Target number: 4 _____

LESSON 3•10

Math Boxes

1. Complete.

   a. Name all the factors of 50.

   _____, _____, _____, _____, _____, _____

   b. Name the factor pairs of 36.

   _____ and _____

   _____ and _____

   _____ and _____

   _____ and _____

   _____ and _____

SRB 7

2. In the 2004 Summer Olympics, which two countries had a combined medal count of 155?

   _____ and _____

Country	Number of Medals
Australia	49
United States	103
China	63
Russia	92

3. What is the mode for the number of books read by the students? Circle the best answer.

   A 3

   B 4

   C 5

   D 2

SRB 73

4. Which of these angles has a measure less than 90 degrees? Circle them.

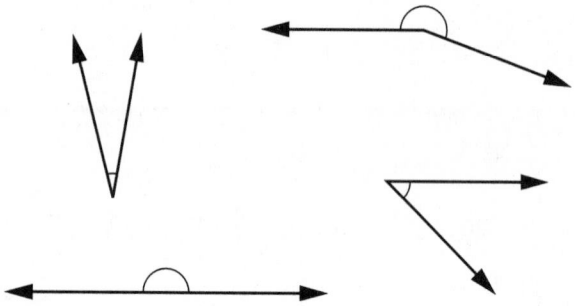

SRB 93

5. a. Measure the line segment to the nearest centimeter.

   About _____ cm

   b. Draw a line segment that is half the length of $\overline{BN}$.

   c. How long is the line segment you drew? About _____ cm

SRB 128

Date Time

LESSON 3•11

Broken Calculator

Solve each open sentence on your calculator without using the "broken" key.
Only one key is broken in each problem. Record your steps.

1.

Broken Key: ⊖

To Solve: $68 + x = 413$

2.

Broken Key: ⊖

To Solve: $z + 643 = 1{,}210$

3.

Broken Key: ⊕

To Solve: $d - 574 = 1{,}437$

4.

Broken Key: ⊗

To Solve: $w / 15 = 8$

Try This

5.

Broken Key: ÷

To Solve: $s * 48 = 2{,}928$

6. Make up one for your partner to solve.

Broken Key: ☐

To Solve:

Date Time

LESSON 3•11

Open Sentences

Solve each open sentence. Copy the entire sentence with the solution in place of the variable. Circle the solution.

1. $48 + d = 70$

   $48 + 22 = 70$

2. $51 = n + 29$

   _____

3. $34 - x = 7$

   _____

4. $32 = 76 - p$

   _____

5. $h - 6 = 9$

   _____

6. $b - 7 = 12$

   _____

7. $u - 30 = 10$

   _____

8. $5 * m = 35$

   _____

9. $y = 3 * 8$

   _____

10. $21 / x = 7$

    _____

11. $x = 32 / 8$

    _____

12. $5 = w / 10$

    _____

Try This

13. Mr. O'Connor wrote two open sentences on the board.

    $45 + x = 71$
    $45 + y = 71$

    Isabel says the two open sentences must have different solutions because the variables are different.

    a. Do you agree with Isabel? _____

    b. Explain your answer.

    _____

    _____

    _____

LESSON 3•11

Estimating Distances

SRB 145

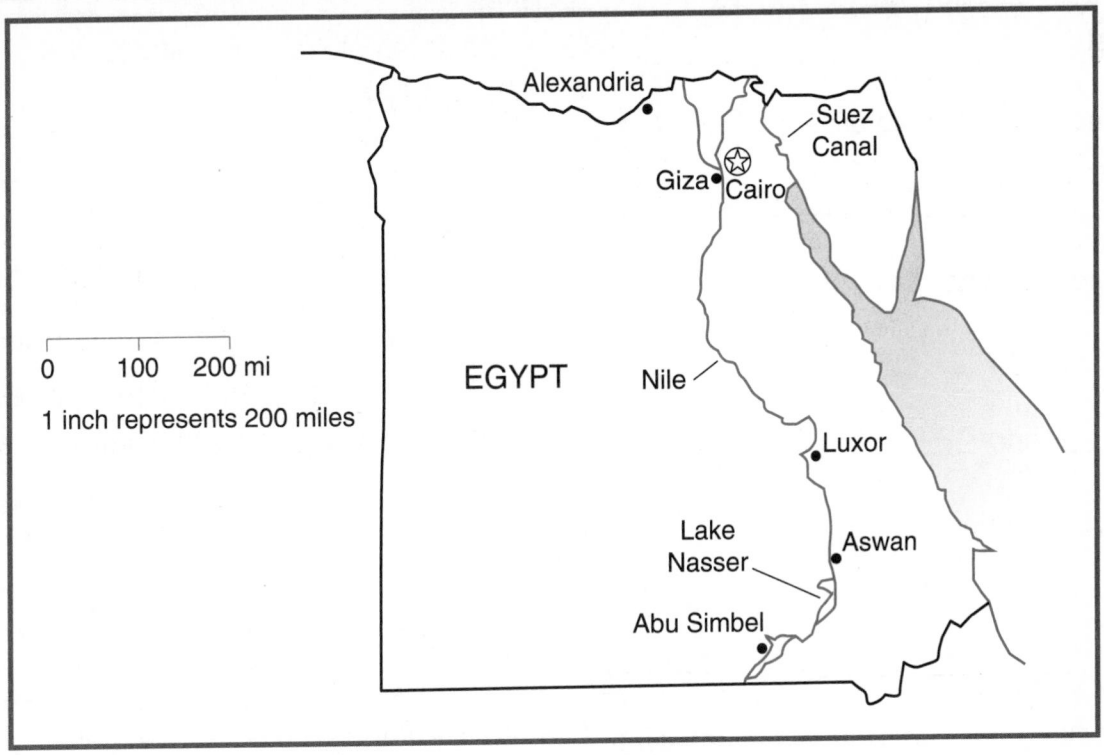

You want to take a trip to Egypt and see the following sights:

- Cairo, the capital, on the Nile River, near the Pyramids at Giza
- Alexandria, a busy modern city and port on the Mediterranean
- The Aswan High Dam across the Nile River, completed in 1970, and Lake Nasser, which formed behind the dam
- The temples at Abu Simbel, built more than 3,000 years ago and moved to their present location in the 1960s to escape the rising water of Lake Nasser

You want to know how far it is between locations.

1. The distance between Alexandria and Abu Simbel is about _____ inch(es) on the map.

   That represents about _____ miles.

2. The distance between Cairo and Aswan is about _____ inch(es) on the map.

   That represents about _____ miles.

3. The distance between Abu Simbel and Aswan is about _____ inch(es) on the map.

   That represents about _____ miles.

Date Time

LESSON 3•11

Math Boxes

1. Complete.

   a. Name all the factors of 12.

   ____, ____, ____, ____, ____, ____

   b. Name the factor pairs of 16.

   ____ and ____

   ____ and ____

   ____ and ____

2. The areas of which two states differ by 944 square miles?

   ____ and ____

State	Total Area
Connecticut	5,543 square miles
Rhode Island	1,545 square miles
Delaware	2,489 square miles
New Jersey	8,721 square miles

3. Use the bar graph to answer the questions.

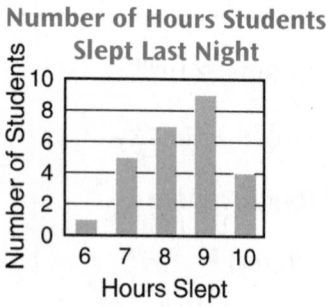

   a. How many students slept 8 hours?

   ____

   b. What is the mode for the number of hours slept?

   ____

4. Which of the angles below have a measure of more than 90 degrees? Circle them.

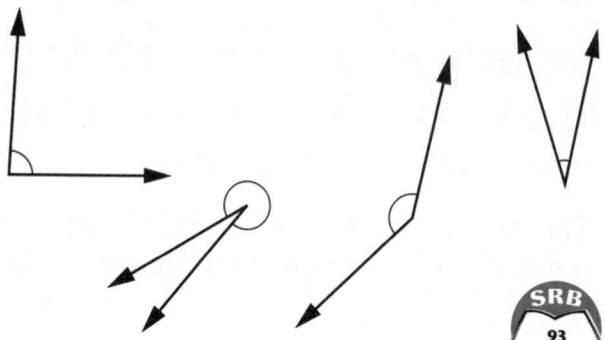

5. a. Measure the line segment to the nearest centimeter.

   About ____ cm

   b. Draw a line segment that is half the length of $\overline{LP}$.

   c. How long is the line segment you drew? About ____ cm

LESSON 3•12

Math Boxes

1. Put these numbers in order from smallest to largest.

0.8 0.08 0.73 0.095

_____

_____

_____

_____

SRB 32

2. Complete.

a. 0.1, 0.9, 1.7, _____, _____, _____

Rule: _____

b. 5.6, 5.3, 5, _____, _____, _____

Rule: _____

c. 7.23, 7.28, 7.33, _____, _____, _____

Rule: _____

SRB 160 161

3. Solve mentally or with a paper-and-pencil algorithm.

a. $11.03
+ $4.79

b. $3.56
− $2.89

SRB 34–37

4. Blake estimated the length of his little finger. Circle the most reasonable estimate.

A 20 millimeters

B 45 millimeters

C 80 millimeters

D 4 millimeters

SRB 130

5. a. Measure the line segment to the nearest centimeter.

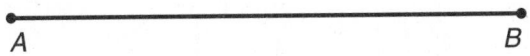

About _____ cm

b. Draw a line segment that is half the length of $\overline{AB}$.

c. How long is the line segment you drew? About _____ cm

SRB 128

Date Time

LESSON 4•1

Place-Value Number Lines

Fill in the missing numbers.

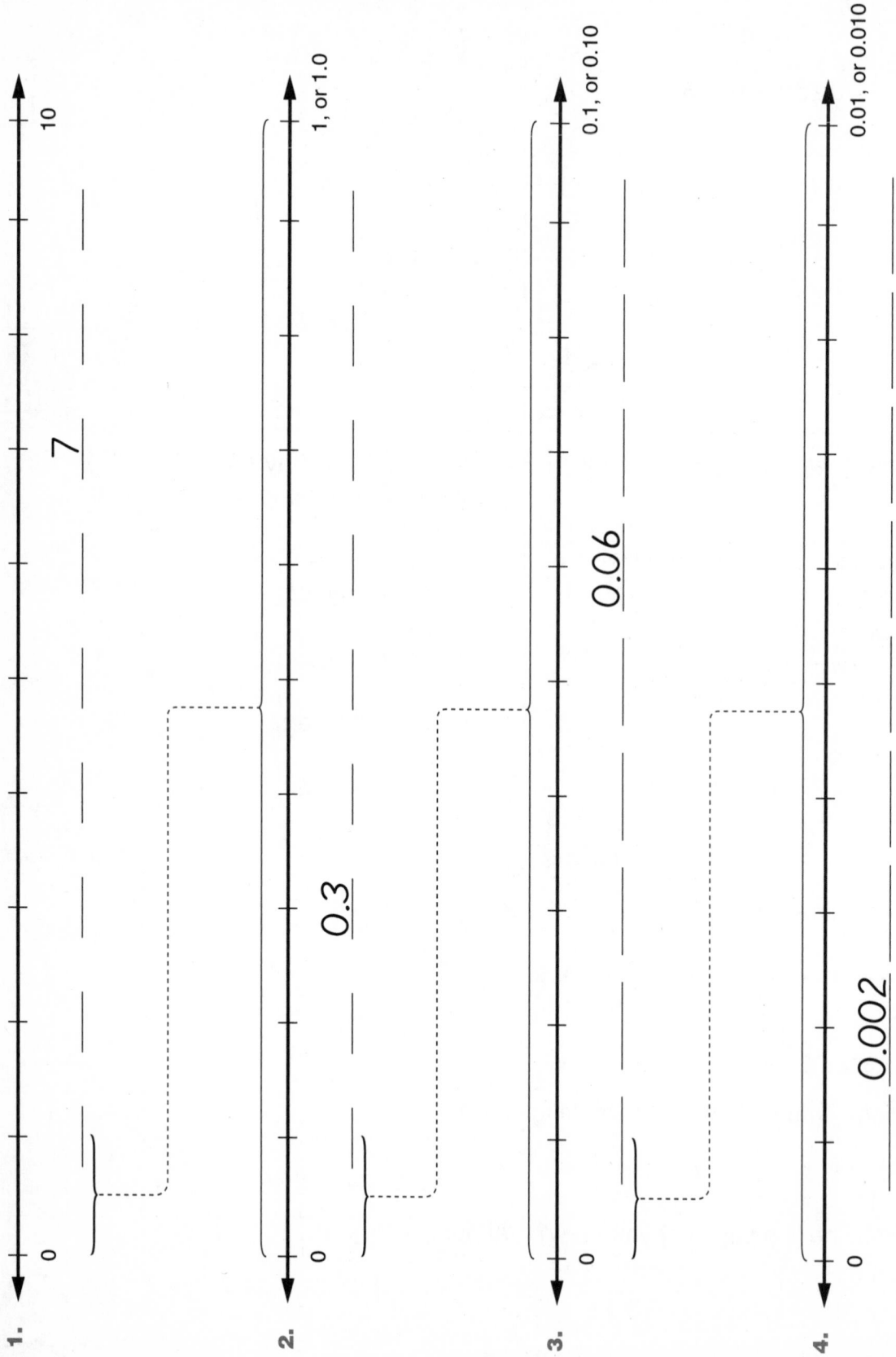

Date Time

LESSON 4•1

Math Boxes

1. Solve mentally.

   a. 9 * 4 = _____

   b. 6 * _____ = 18

   c. 3 * _____ = 21

   d. 16 ÷ 4 = _____

   e. 20 ÷ 4 = _____

   f. 54 ÷ 6 = _____

SRB 16 20

2. Solve 199 = p − 408.
   Choose the best answer.

   ⬭ p = 209

   ⬭ p = 309

   ⬭ p = 607

   ⬭ p = 507

SRB 148

3. In the numeral 9,358,461.72, the 6 is worth 60.

   a. The 4 is worth _____.

   b. The 8 is worth _____.

   c. The 3 is worth _____.

   d. The 9 is worth _____.

   e. The 7 is worth _____.

SRB 30 31

4. Draw and label ray *BY*.
   Draw point *A* on it.

SRB 91

5. Insert parentheses to make these number sentences true.

   a. 5 * 4 − 2 = 18

   b. 25 + 8 * 7 = 81

   c. 1 = 36 / 6 − 5

   d. 19 = 15 − 5 + 81 / 9

SRB 150

6. Estimate the sum. Write a number model to show how you estimated.

   458 + 1,999 + 12,307

   Number model: _____

   _____

   _____

SRB 181

Date Time

LESSON 4•2

Tenths and Hundredths

Base-10 Block	Symbol	Value
Flat	□	1
Long	|	$\frac{1}{10}$, or 0.1
Cube	▪	$\frac{1}{100}$, or 0.01

SRB 27 28

1. Complete the table.

Base-10 Blocks	Fraction Notation	Decimal Notation
||	$\frac{2}{10}$	0.2
▪▪▪▪▪		
||||| ||||		
□ ||| ▪▪▪▪▪		

2. Write each number in decimal notation.

Example: $\frac{3}{10}$ = 0.3

a. $\frac{4}{10}$ = _____ b. $\frac{71}{100}$ = _____ c. $32\frac{6}{100}$ = _____

3. Draw base-10 blocks to show each number. Draw as few blocks as possible.

Example: 0.3 |||

a. 0.43 b. 2.16 c. 0.07

4. Write each of the following in decimal notation.

a. 8 tenths _____ b. 82 hundredths _____ c. 38 and 5 tenths _____

Try This

5. Write 53 thousandths in decimal notation. _____

Date Time

LESSON 4•2

Math Boxes

1. a. What is the maximum number of blocks any student lives from school?

_____

b. What is the minimum number of blocks?

_____

c. What is the mode? _____

d. What is the median number of blocks?

_____

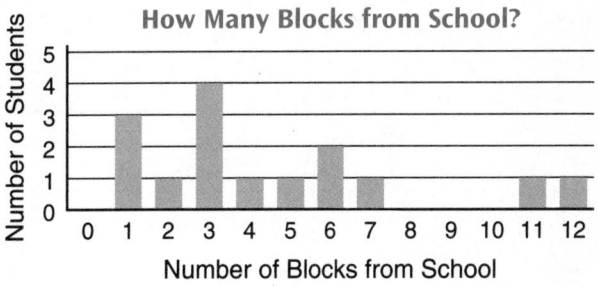

SRB 73

2. Solve mentally or with a paper-and-pencil algorithm.

a. $\begin{array}{r} 391 \\ +\ 467 \\ \hline \end{array}$

b. $\begin{array}{r} 983 \\ -\ 494 \\ \hline \end{array}$

SRB 10–15

3. If 1 inch on a map represents 40 miles, then

a. 2 in. represent _____ mi.

b. 4 in. represent _____ mi.

c. 5 in. represent _____ mi.

d. $2\frac{1}{2}$ in. represent _____ mi.

e. $1\frac{3}{4}$ in. represent _____ mi.

SRB 145

4. Write as dollars and cents.

a. 20 dimes = $_____._____

b. 20 nickels = $_____._____

c. 20 quarters = $_____._____

d. 10 quarters and 7 dimes = $_____._____

5. Solve mentally.

a. $5 * 3 =$ _____

b. $5 * 30 =$ _____

c. _____ $= 5 * 9$

d. _____ $= 50 * 9$

e. $6 * 7 =$ _____

f. $60 * 70 =$ _____

SRB 17

Date Time

LESSON 4•3

Comparing Decimals

Math Message

1. Arjun thought that 0.3 was less than 0.15. Explain or draw pictures to help Arjun see that 0.3 is more than 0.15.

_____

_____

_____

_____

_____

2. Use base-10 blocks to complete the following table.

"<" means "is less than."

">" means "is greater than."

Base-10 Blocks	Decimal	>, <, or =	Decimal	Base-10 Blocks
‖	0.2	>	0.12	|▪▪
▪▪▪▪▪			0.1	
	0.13			‖|▪
‖|▪▪▪			0.3	
	1.2			□□|
‖‖▪▪▪▪▪			0.39	
	2.3			□‖‖‖‖ ‖‖‖‖ ‖‖‖

Date Time

LESSON 4•3 Ordering Decimals

1. Write < or >.

a. 0.24 _____ 0.18 b. 0.05 _____ 0.1 c. 0.2 _____ 0.35

d. 1.03 _____ 0.30 e. 3.2 _____ 6.59 f. 25.9 _____ 25.72

2. Write your own decimals to make true number sentences.

a. _____ > _____ b. _____ < _____ c. _____ < _____

3. Put these numbers in order from smallest to largest.

a. 0.05, 0.5, 0.55, 5.5 _____ _____ _____ _____
smallest largest

b. 0.99, 0.27, 1.8, 2.01 _____ _____ _____ _____
smallest largest

c. 2.1, 2.01, 20.1, 20.01 _____ _____ _____ _____
smallest largest

d. 0.01, 0.10, 0.11, 0.09 _____ _____ _____ _____
smallest largest

4. Write your own decimals in order from smallest to largest.

_____ _____ _____ _____
smallest largest

5. "What's green inside, white outside, and hops?"
To find the answer, put the numbers in order from smallest to largest.

0.66	1	0.2	1.05	0.90	0.01	0.75	0.35	$\frac{25}{100}$	$\frac{50}{100}$	0.05	0.09	5.5
N	I	O	C	W	A	D	S	G	A	F	R	H

Write your answers in the following table. The first answer is done for you.

0.01												
A												

LESSON 4•3

Math Boxes

1. Solve mentally.

a. 5 ∗ _____ = 40

b. 9 ∗ 9 = _____

c. 9 ∗ _____ = 27

d. 42 ÷ 6 = _____

e. 54 ÷ 9 = _____

f. 63 ÷ 7 = _____

SRB 16 20

2. Solve each open sentence.

a. 100 + w = 175 w = _____

b. 503 + y = 642 y = _____

c. p + 263 = 319 p = _____

d. 444 − s = 93 s = _____

e. r − 320 = 600 r = _____

SRB 148

3. In 34.561

a. The 3 is worth _____.

b. The 4 is worth _____.

c. The 5 is worth _____.

d. The 6 is worth _____.

e. The 1 is worth _____.

SRB 30 31

4. Draw and label ray *CT*.
Draw point *A* on it.

SRB 91

5. Insert parentheses to make these number sentences true.

a. 7 ∗ 8 − 6 = 50

b. 13 − 4 ∗ 6 = 54

c. 10 = 3 + 49 / 7

d. 28 = 28 − 6 + 42 / 7

SRB 150

6. Estimate the sum. Write a number model to show how you estimated.

3,005 + 9,865 + 2,109

Number model: _____

_____

_____

SRB 181

Date Time

LESSON 4•4

A Bicycle Trip

Diego and Alex often take all-day bicycle trips together. During the summer, they took a 3-day bicycle tour. They carried camping gear in their saddlebags for the two nights they would be away from home.

Alex had a **trip meter** that showed miles traveled in tenths of miles. He kept a log of the distances they traveled each day before and after lunch.

Travel Log

	Distance Traveled	
Timetable	Before lunch	After lunch
Day 1	27.0 mi	31.3 mi
Day 2	36.6 mi	20.9 mi
Day 3	25.8 mi	27.0 mi

Use estimation to answer the following questions. Do not work the problems out on paper or with a calculator.

1. On which day did they travel the most miles? _____

2. On which day did they travel the fewest miles? _____

3. During the whole trip, did they travel more miles before or after lunch? _____

4. Estimate the total distance they traveled. Choose the best answer.

   - ⬭ less than 150 miles
   - ⬭ between 150 and 180 miles
   - ⬭ between 180 and 200 miles
   - ⬭ more than 200 miles

5. Explain how you solved Problem 4.

   _____

   _____

   _____

6. On Day 1, about how many more miles did they travel after lunch than before lunch?

   _____

7. Diego said that they traveled 1.2 more miles before lunch on Day 1 than on Day 3. Alex disagreed. He said they traveled 2.2 more miles. Who is right? Explain your answer.

   _____

   _____

Date Time

LESSON 4•4

Math Boxes

1. a. What is the maximum number of movies a student viewed in a month?

_____

b. What is the minimum number of movies? _____

c. What is the mode? _____

d. What is the median? _____

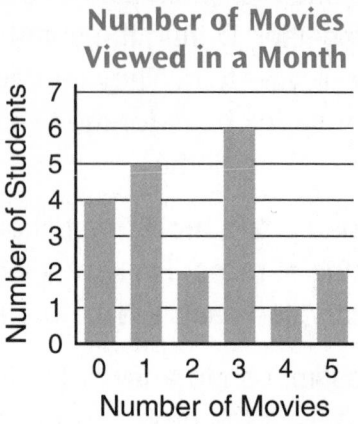

SRB 73

2. Solve mentally or with a paper-and-pencil algorithm.

a. 814 + 123

b. 754 − 396

SRB 10–15

3. If 2 centimeters on a map represent 50 kilometers, then

a. 1 cm represents _____ km.

b. 3 cm represent _____ km.

c. 4 cm represent _____ km.

d. 0.5 cm represent _____ km.

e. 8.5 cm represent _____ km.

SRB 145

4. Write 40 quarters and 3 dimes in dollars-and-cents notation. Choose the best answer.

- ⬭ $4.03
- ⬭ $4.30
- ⬭ $10.30
- ⬭ $40.30

5. Solve mentally.

a. 6 ∗ 9 = _____

b. 6 ∗ 90 = _____

c. _____ = 5 ∗ 8

d. _____ = 50 ∗ 8

e. 4 ∗ 4 = _____

f. 4 ∗ 40 = _____

SRB 17

Date Time

LESSON 4•5

Decimal Addition and Subtraction

Add or subtract mentally or with a paper-and-pencil algorithm.
Pay attention to the + and − symbols.

SRB 34–37

1. 2.05 + 1.83 = _____
2. 3.04 + 2.8 = _____
3. 2.4 + 3.01 + 0.26 = _____
4. 2.31 − 1.88 = _____
5. 19 + 1.9 = _____
6. 1 − 0.67 = _____

7. Choose one of the problems from above. Explain the method you used to solve the problem.

_____

_____

_____

Date Time

LESSON 4•5 Circle Graphs

Percent urban is the number of people out of 100 who live in towns or cities. *Percent rural* is the number of people out of 100 who live in the countryside. Each circle graph below represents the percent of the urban and rural population of an African country.

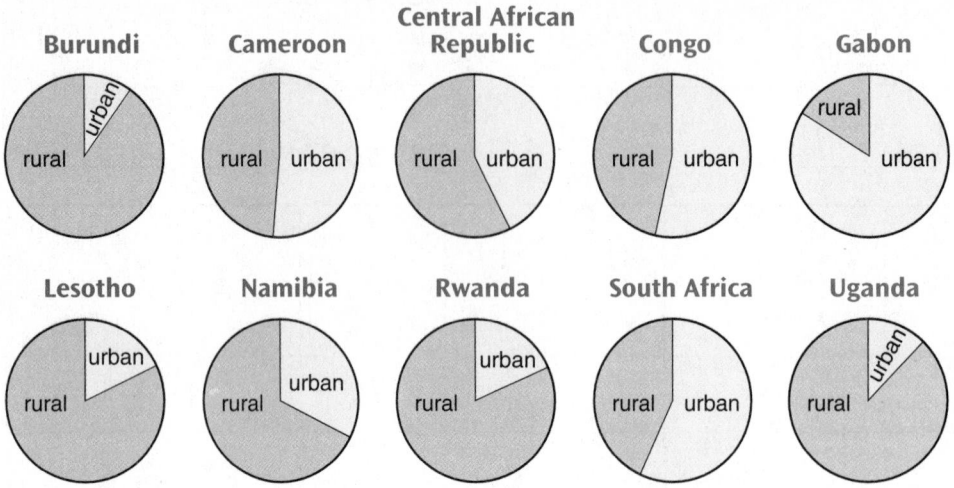

Source: The United Nations

1. For each pair, circle the country with the larger urban population.

a. Congo Uganda **b.** Rwanda Gabon

c. Burundi South Africa **d.** Namibia Lesotho

2. Which country has the greatest percentage of people living in *urban* areas? _____

3. Which two countries have the greatest percentage of people living in *rural* areas? _____

4. Which two countries have about $\frac{1}{2}$ of their people living in urban areas and $\frac{1}{2}$ of their people living in rural areas? _____

Try This

5. Write a question that can be answered from the information in the graphs. Then answer the question.

Question: _____

_____

Answer: _____

LESSON 4•5

Math Boxes

1. Insert >, <, or =.

   a. 0.96 _____ 0.4

   b. 0.50 _____ 0.500

   c. 1.3 _____ 1.09

   d. 0.85 _____ 0.86

   e. 0.700 _____ 0.007

SRB 32 33

2. a. Measure the length of this line segment to the nearest $\frac{1}{2}$ centimeter.

   _____

   About _____ cm

   b. Draw a line segment 3 centimeters long.

SRB 128

3. Fill in the missing numbers.

   a. 28, 35, 42, _____, _____, _____

   Rule: _____

   b. 56, 48, 40, _____, _____, _____

   Rule: _____

   c. 81, _____, 63, _____, 45, _____

   Rule: _____

SRB 160 161

4. Solve each open sentence.

   a. $5.9 - T = 5$ $T =$ _____

   b. $9.4 - K = 3$ $K =$ _____

   c. $0.81 - M = 0.43$ $M =$ _____

   d. $F - 2.1 = 6.8$ $F =$ _____

   e. $2.43 = S + 1.06$ $S =$ _____

   f. $R - 12.2 = 4.65$ $R =$ _____

SRB 148

5. Add 9 tens, 8 hundredths, and 3 tenths to 34.53.

   What is the result? _____

SRB 36

6. Add mentally or with a paper-and-pencil algorithm.

   a.
   $$\begin{array}{r} 6 \\ 40 \\ 150 \\ +\ 1{,}000 \\ \hline \end{array}$$

   b.
   $$\begin{array}{r} 54 \\ 180 \\ 240 \\ +\ 800 \\ \hline \end{array}$$

SRB 10 11

LESSON 4•6

Keeping a Bank Balance

SRB 34–37

Math Message

Solve. Show your work on the grid.

1. Cleo went to the store to buy school supplies. She bought a notebook for $2.39, a pen for $0.99, and a set of markers for $3.99. How much money did she spend in all?

   _____

2. Nicholas went to the store with a $20 bill. His groceries cost $13.52. How much change did he get?

   _____

On January 2, Kate's aunt opened a bank account for Kate. Her aunt deposited $100.00 in the account.

Over the next several months, Kate made regular deposits into her account. She deposited part of her allowance and most of the money she made babysitting.

Kate also made a few withdrawals—to buy a radio and some new clothes.

Think about the answers to the following questions:

- When you **withdraw** money, do you take money out or put money in?
- When you **deposit** money, do you take money out or put money in?
- When your money earns **interest,** does this add money to your account or take money away?

The table on the next page shows the transactions (deposits and withdrawals) that Kate made during the first 4 months of the year and the interest she earned.

Date Time

LESSON 4•6

Keeping a Bank Balance *continued*

3. In March, Kate took more money out of her bank account than she put in. In which other month did she withdraw more money than she deposited? _____

4. Estimate whether Kate will have more or less than $100.00 at the end of April. _____

5. Complete the table. Remember to add if Kate makes a deposit or earns interest and to subtract if she makes a withdrawal.

Date	Transaction		Current Balance
January 2	Deposit	$100.00	$ 100.00
January 14	Deposit	$14.23	+ $ 14.23 $ 114.23
February 4	Withdrawal	$16.50	$ _____ $ _____
February 11	Deposit	$33.75	$ _____ $ _____
February 14	Withdrawal	$16.50	$ _____ $ _____
March 19	Deposit	$62.00	$ _____ $ _____
March 30	Withdrawal	$104.26	$ _____ $ _____
March 31	Interest	$0.78	$ _____ $ _____
April 1	Deposit	$70.60	$ _____ $ _____
April 3	Withdrawal	$45.52	$ _____ $ _____
April 28	Withdrawal	$27.91	$ _____ $ _____

LESSON 4•6

Math Boxes

1. Solve mentally or with a paper-and-pencil algorithm.

   a. $5.18 − $3.65 = _____

   b. $16.86 + $9.24 = _____

   c. 0.87 + 0.94 = _____

   d. 11.2 − 3.9 = _____

SRB 34–37

2. Put these numbers in order from smallest to largest.

   5.92 0.95 9.25 2.95 0.92

   _____ _____ _____ _____ _____

SRB 32 33

3. A trumpeter swan can weigh about 16.8 kilograms. A Manchurian crane can weigh about 14.9 kilograms. How much heavier is a trumpeter swan than a Manchurian crane?

   _____ kilograms

4. Number of items students brought to the school food drive:

   28, 26, 3, 8, 2, 6, 8, 13, 1, 5

   What is the

   a. maximum? ____ b. minimum? ____

   c. range? ____ d. mode? ____

   e. median? ____ f. mean? ____

SRB 73 75

5. How do you write the following number using digits: six-hundred million, five thousand, twenty-one? Choose the best answer.

   ⬭ 6,005,210

   ⬭ 600,500,021

   ⬭ 600,500,210

   ⬭ 600,005,021

SRB 4

Date Time

LESSON 4•7

Math Boxes

1. Insert >, <, or =.

   a. 0.6 _____ 0.57

   b. 0.37 _____ 0.36

   c. 2.56 _____ 2.056

   d. 0.24 _____ 0.240

   e. 0.008 _____ 0.080

SRB 32 33

2. a. Measure the length of this line segment to the nearest $\frac{1}{2}$ centimeter.

   _____

   About _____ cm

   b. Draw a line segment 7 centimeters long.

SRB 128

3. Fill in the missing numbers.

   a. 16, 20, 24, _____, _____, _____

   Rule: _____

   b. 15, _____, 21, _____, 27, _____

   Rule: _____

   c. _____, _____, 30, _____, _____, 60

   Rule: _____

SRB 160 161

4. Solve each open sentence.

   a. $D + 1.0 = 9.2$ $D =$ _____

   b. $5.6 + K = 10$ $K =$ _____

   c. $0.8 - M = 0.6$ $M =$ _____

   d. $9.09 + S = 13.64$ $S =$ _____

   e. $F + 1.25 = 12.90$ $F =$ _____

   f. $R - 0.03 = 1.65$ $R =$ _____

SRB 148

5. Add 6 tens, 4 hundredths, and 2 tenths to 367.53.

   What is the result? _____

SRB 36

6. Add mentally or with a paper-and-pencil algorithm.

   a.
   $$\begin{array}{r} 24 \\ 80 \\ 360 \\ \underline{+\ 1{,}200} \end{array}$$

   b.
   $$\begin{array}{r} 6 \\ 180 \\ 40 \\ \underline{+\ 1{,}200} \end{array}$$

SRB 10 11

Date Time

LESSON 4•7

Tenths, Hundredths, and Thousandths

Math Message

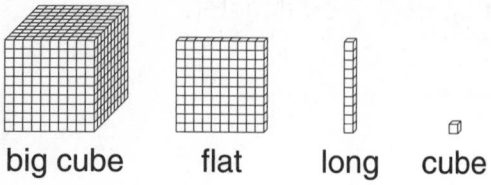

1. 1 big cube = _____ flats

   1 flat = 1/10 of a big cube

2. 1 big cube = _____ longs

   1 long = 1/[ ] of a big cube

3. 1 big cube = _____ cubes

   1 cube = 1/[ ] of a big cube

Base-10 Block	Symbol	Value
Big Cube		1
Flat		$\frac{1}{10}$, or 0.1
Long		$\frac{1}{100}$, or 0.01
Cube		$\frac{1}{1,000}$, or 0.001

4. Complete the table.

Base-10 Blocks	Fraction Notation	Decimal Notation
	$\frac{142}{1,000}$	0.142

LESSON 4•7 Tenths, Hundredths, and Thousandths *cont.*

5. Write each number in decimal notation.

SRB 30 31

Example: $\frac{72}{1{,}000}$ = 0.072

a. $\frac{416}{1{,}000}$ = _____ **b.** $\frac{17}{1{,}000}$ = _____ **c.** $\frac{8}{1{,}000}$ = _____

d. $7\frac{9}{100}$ = _____ **e.** $8\frac{14}{1{,}000}$ = _____ **f.** $385\frac{4}{10}$ = _____

6. Draw base-10 blocks to show each number.

Example: 0.205

a. 0.21 **b.** 0.306

c. 0.008 **d.** 1.054

7. Write each of the following in decimal notation.

a. 284 thousandths _____ **b.** 36 hundredths _____

c. 7 thousandths _____ **d.** 90 and 16 thousandths _____

e. 15 and 3 tenths _____ **f.** 408 thousandths _____

8. Write < or >.

a. 0.302 _____ 0.203 **b.** 0.51 _____ 0.310 **c.** 0.816 _____ 0.9

d. 1.47 _____ 1.5 **e.** 0.073 _____ 0.73 **f.** 4.01 _____ 4.009

Try This

9. Put these numbers in order from smallest to largest: 0.03, 0.009, 0.285, 0.064

_____ _____ _____ _____

smallest largest

Date Time

LESSON 4•8

Measuring Length with Metric Units

1. Your teacher will choose several objects or distances to measure. Measure each to the nearest centimeter. Then compare your measurements with your partner's. If you do not agree, work together to measure the objects again. Record the results in the table.

Object or Distance	My Measurement	Partner's Measurement	Agreed Measurement
	About _____ cm	About _____ cm	About _____ cm
	About _____ cm	About _____ cm	About _____ cm
	About _____ cm	About _____ cm	About _____ cm
	About _____ cm	About _____ cm	About _____ cm
	About _____ cm	About _____ cm	About _____ cm

2. Measure these line segments to the nearest centimeter.

   a. _____

   About _____ centimeters

   b. _____

   About _____ centimeters

3. Measure these line segments to the nearest $\frac{1}{2}$ centimeter.

   a. _____

   About _____ centimeters

   b. _____

   About _____ centimeters

Date Time

LESSON 4•8

Math Boxes

1. Solve mentally or with a paper-and-pencil algorithm.

   a. 3,309 + 721

   b. 2,700 − 1,299

SRB 10–15

2. Complete.

   a. 1 cm = _____ mm

   b. 5 cm = _____ mm

   c. _____ cm = 30 mm

   d. 100 cm = _____ mm

   e. 200 cm = _____ mm

SRB 129

3. Tell whether each number sentence is true or false.

   a. $8.77 - 0.08 = 8.50$ _____

   b. $35.7 + 22.1 = 57.87$ _____

   c. $90.2 - 44.9 < 45$ _____

   d. $4.66 + 2.13 > 6$ _____

SRB 36 37 148

4. Trace at least two regular polygons from your Geometry Template.

SRB 97

5. Without measuring, estimate the length of your foot from heel to toe. Then measure the length of your foot.

   a. Estimate:

   About _____ cm

   b. Measurement:

   About _____ cm

SRB 128 130

6. Complete.

   a. Is 47 closer to 40 or 50?

   _____

   b. Name the number halfway between 30 and 40.

   _____

SRB 182 183

Date Time

LESSON 4•9 Personal References for Units of Length

Personal References for Metric Units of Length

SRB 130

Use a ruler, meterstick, or tape measure to find common objects that have lengths of 1 centimeter, 1 decimeter, and 1 meter. The lengths do not have to be exact, but they should be close. Ask a friend to look for references with you. You can find more than one reference for each unit. Record the references in the table below.

Unit of Measure	Personal References
1 centimeter (cm)	
1 decimeter (dm), or 10 centimeters	
1 meter (m)	

To be completed in Lesson 5-1.

Personal References for U.S. Customary Units of Length

Use a ruler, yardstick, or tape measure to find common objects that have lengths of 1 inch, 1 foot, and 1 yard. The lengths do not have to be exact, but they should be close. Ask a friend to look for references with you. You can find more than one reference for each unit. Record the references in the table below.

Unit of Measure	Personal References
1 inch (in.)	
1 foot (ft)	
1 yard (yd)	

Date Time

LESSON 4•9

Measurement Collection for Metric Units of Length

Use your personal references to estimate the length of an object or a distance in centimeters, decimeters, or meters. Describe the object or distance and record your estimate in the table below. Then measure the object or distance and record the actual measurement in the table.

SRB 130

Object or Distance	Estimated Length	Actual Length

LESSON 4•9

Math Boxes

1. Solve mentally or with a paper-and-pencil algorithm.

   a. $12.63 + $5.66 = _____

   b. $2.46 − $1.34 = _____

   c. 9.6 − 4.8 = _____

   d. 0.64 + 0.47 = _____

SRB 34–37

2. Put these numbers in order from smallest to largest.

   1.68 0.78 6.71 0.61

   _____ _____ _____ _____

SRB 32 33

3. The great spotted kiwi bird is about 114.3 centimeters tall. The greater rhea is about 137.1 centimeters tall. How much taller is the greater rhea than the great spotted kiwi bird?

   _____ centimeters

4. Make up a set of 7 numbers having the following landmarks:

   mode: 21
   median: 24
   maximum: 35
   range: 20

   ___, ___, ___, ___, ___, ___, ___

SRB 73

5. Write the following numbers using digits:

   a. four hundred eighty-two thousand, one hundred ninety-seven

   _____

   b. eight hundred million, twelve thousand, five

   _____

SRB 4

Date Time

LESSON 4•10

Measuring in Millimeters

Math Message

On your centimeter ruler, the numbered marks are for centimeters and the little marks between the centimeter marks are for millimeters.

1. Look at your centimeter ruler. How many millimeters are in 1 centimeter? _____ mm

2. Name something that measures about 1 millimeter. _____

_____

_____

3. Draw a line segment that is 8 centimeters long.

4. Draw a line segment that is 80 millimeters long.

Measure each line segment below using both the millimeter side and the centimeter side of the cm/mm ruler. Record both measurements.

5. *A* _____ *B*

Length of $\overline{AB}$ = _____ mm = _____ cm

6. *C* _____ *D* Length of $\overline{CD}$ = _____ mm = _____ cm

7. *E* ___ *F* Length of $\overline{EF}$ = _____ mm = _____ cm

---

Measuring Land Invertebrates

An invertebrate is an animal that does not have a backbone. (The backbone is also called the spinal column.) Some invertebrates live on land, others in water. The most common land invertebrates are insects.

The invertebrates shown on page 102, except the earthworm, bumblebee, and mealybug, have been drawn to about actual size. The earthworm can grow to about 4 times the length shown. The bumblebee is shown about twice its actual size and the mealybug about 3 times its actual size.

Date Time

LESSON 4•10

Measuring Land Invertebrates

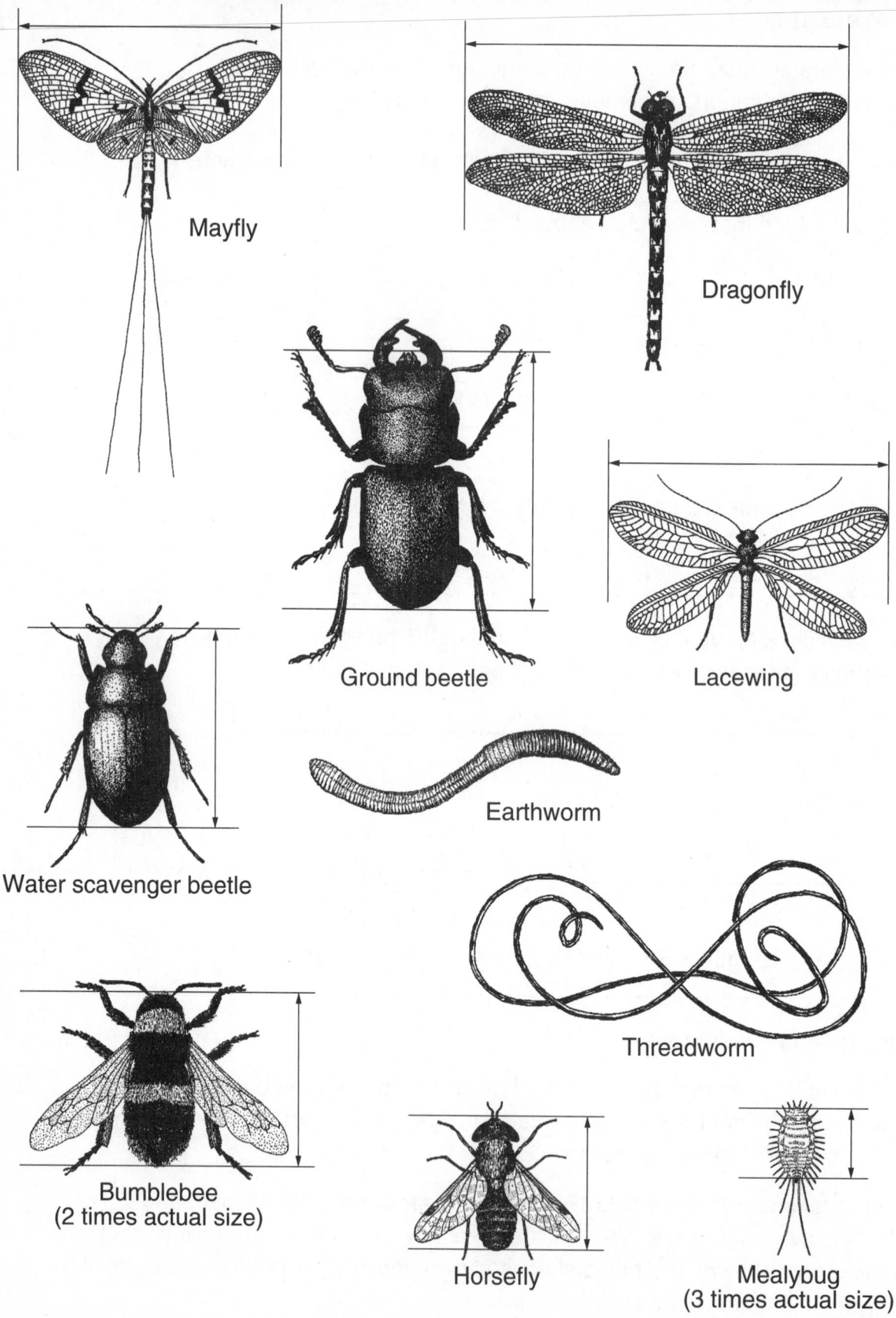

LESSON 4•10 Measuring Land Invertebrates *continued*

Refer to the pictures on page 102 to answer the following questions.

1 centimeter (cm) = 10 millimeters (mm)
1 millimeter = 0.1 centimeter

1. Measure the following invertebrates to the nearest millimeter by finding the distance between the two guidelines. Then give the lengths in centimeters.

a. mayfly	About _____ mm	About _____ cm
b. dragonfly	About _____ mm	About _____ cm
c. water scavenger beetle	About _____ mm	About _____ cm
d. ground beetle	About _____ mm	About _____ cm
e. lacewing	About _____ mm	About _____ cm
f. horsefly	About _____ mm	About _____ cm

2. How much longer is the ground beetle than the water scavenger beetle? About _____ cm

3. The bee has been drawn to twice its actual size.
In reality, which is longer, the bee or the horsefly? _____

How much longer? About _____ mm

4. The mealybug has been drawn to 3 times its actual size. In the space at the right, draw a mealybug that is about the actual size.

5. What is the actual size of the mealybug in millimeters? _____ mm

6. How did you solve Problem 5?

_____

_____

_____

7. When straight, the threadworm in the drawing is 306 millimeters long.

What is its length in centimeters? _____ cm In meters? _____ m

Date Time

LESSON 4•10

Math Boxes

1. Solve mentally or with a paper-and-pencil algorithm.

   a. 4,647 + 3,228

   b. 2,500 − 1,398

SRB 10–15

2. Complete.

   a. 7 cm = _____ mm

   b. 15 cm = _____ mm

   c. 500 cm = _____ m

   d. _____ cm = 40 mm

   e. _____ cm = 8 m

SRB 129

3. Tell whether each number sentence is true or false.

   a. 2.34 − 0.09 = 2.25 _____

   b. 89.6 + 21.7 = 111.3 _____

   c. 56.4 − 23.8 < 33 _____

   d. 5.17 + 3.86 > 10 _____

SRB 36 37 148

4. Name two properties of a regular polygon.

   a. _____

   _____

   b. _____

   _____

SRB 97

5. Without measuring, estimate the height of your chair. Then measure it.

   a. Estimate:

   About _____ cm

   b. Measurement:

   About _____ cm

SRB 128 130

6. Complete.

   a. Is 326 closer to 300 or 400?

   _____

   b. Name the number halfway between 500 and 800.

   _____

SRB 182 183

Date Time

LESSON 4•11

Math Boxes

1. Estimate the sum. Write a number model to show how you estimated.

   3,721 + 2,876 + 7,103

   Number model: _____

   _____

   _____

SRB 181

2. Solve mentally.

   a. 4 * 8 = _____

   b. 4 * 80 = _____

   c. _____ = 5 * 3

   d. _____ = 50 * 3

   e. 6 * 6 = _____

   f. 6 * 60 = _____

SRB 16 17

3. Complete.

   a. Is 63 closer to 60 or 70? _____

   b. What number is halfway between 80 and 90? _____

   c. Is 572 closer to 500 or 600? _____

   d. What number is halfway between 300 and 600? _____

SRB 182 183

4. Write the following numbers using digits:

   a. one million, three hundred forty-six thousand, thirteen

   _____

   b. twenty-two million, fifteen thousand, three hundred fifty-four

   _____

SRB 4

5. Add mentally or with a paper-and-pencil algorithm.

a.	b.	c.	d.
35	18	54	48
100	420	180	720
280	120	360	180
+ 800	+ 2,800	+ 1,200	+ 2,700

SRB 10 11

Date Time

LESSON 5•1 Multiplying Ones by Tens

You can extend a multiplication fact by making one of the factors a multiple of ten.

Example:

Original fact: 2 ∗ 3 = 6

Extended facts: 2 ∗ 30 = _____, or 20 ∗ 3 = _____

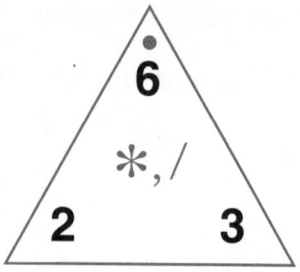

Write a multiplication fact for each Fact Triangle shown below.
Then extend this fact by changing one factor to a multiple of ten.

1.

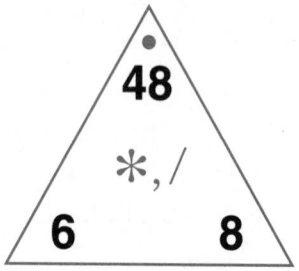

Original fact: _____

Extended fact: _____

2.

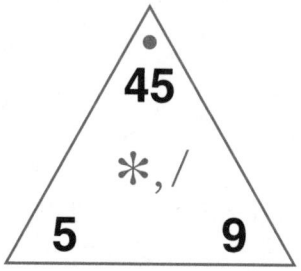

Original fact: _____

Extended fact: _____

3.

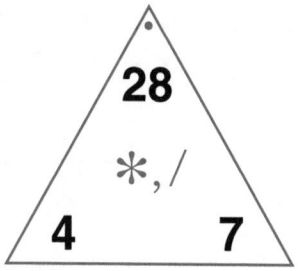

Original fact: _____

Extended fact: _____

4.

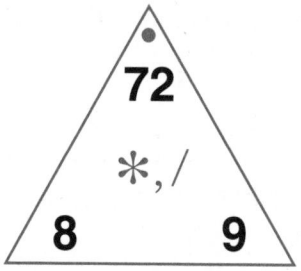

Original fact: _____

Extended fact: _____

5. What shortcut can you use to multiply ones by tens, such as 3 ∗ 60?

_____

_____

Date Time

LESSON 5•1

Multiplying Tens by Tens

You can extend a multiplication fact by making both factors multiples of ten.

SRB 17

Example:

Original fact: 3 ∗ 5 = 15

Extended fact: 30 ∗ 50 = _____

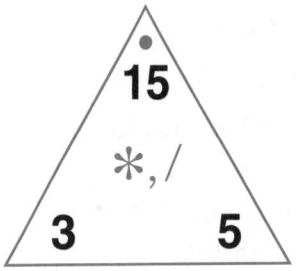

Write a multiplication fact for each Fact Triangle shown below.
Then extend this fact by changing both factors to multiples of ten.

1.

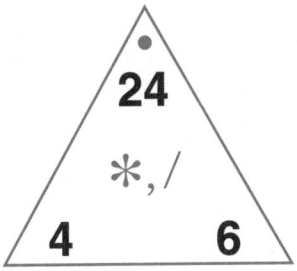

Original fact: _____

Extended fact: _____

2.

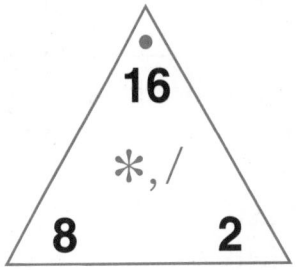

Original fact: _____

Extended fact: _____

3.

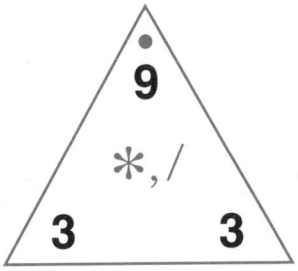

Original fact: _____

Extended fact: _____

4.

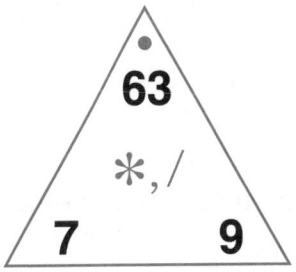

Original fact: _____

Extended fact: _____

5. What shortcut can you use to multiply tens by tens, such as 40 ∗ 60?

_____

_____

Date Time

LESSON 5•1

Math Boxes

1. A number has

   6 in the tenths place,
   9 in the hundreds place,
   2 in the thousands place,
   7 in the ones place,
   3 in the tens place, and
   5 in the hundredths place.

   Write the number.

   ____, ____ ____ ____ . ____ ____

   SRB 31

2. Solve mentally.

   a. 40 * 50 = _____

   b. 70 * 300 = _____

   c. 60 * _____ = 180

   d. 90 * _____ = 810

   e. _____ * 9 = 7,200

   SRB 17

3. Solve mentally or with a paper-and-pencil algorithm.

   a.
   ```
     4,500
       540
       100
   +    12
   ------
   ```

   b.
   ```
     2,100
       420
        90
   +    18
   ------
   ```

   SRB 10 11

4. List all the factors of these numbers.

   a. 40 _____

   _____

   b. 18 _____

   _____

   c. 28 _____

   _____

   SRB 7

5. If 1 inch on a map represents 300 miles, then

   a. 6 in. → _____ miles

   b. 10 in. → _____ miles

   c. _____ in. → 900 miles

   d. _____ in. → 750 miles

   e. $8\frac{1}{2}$ in. → _____ miles

   SRB 145

6. a. Five children share 27 tennis balls equally.

   Each child gets _____ balls.

   There are _____ balls left over.

   b. There are 32 cookies for 6 friends.

   Each friend gets _____ cookies.

   There are _____ cookies left over.

   SRB 20

LESSON 5•2 Presidential Information

The following table shows the dates on which the most recent presidents of the United States were sworn in and their ages at the time they were sworn in.

President	Date Sworn In	Age
F. D. Roosevelt	March 4, 1933	51
Truman	April 12, 1945	60
Eisenhower	January 20, 1953	62
Kennedy	January 20, 1961	43
Johnson	November 22, 1963	55
Nixon	January 20, 1969	56
Ford	August 9, 1974	61
Carter	January 20, 1977	52
Reagan	January 20, 1981	69
G. H. Bush	January 20, 1989	64
Clinton	January 20, 1993	46
G. W. Bush	January 20, 2001	54

1. What is the median age (the middle age) of the presidents at the time they were sworn in? _____ years

2. What is the range of their ages (the difference between the ages of the oldest and the youngest)? _____ years

3. Who was president for the longest time? _____

4. Who was president for the shortest time? _____

5. Presidents are elected to serve for 1 term. A term lasts 4 years. Which presidents served only 1 term or less than 1 term?

   _____

6. Which president was sworn in about 28 years after Roosevelt? _____

7. Roosevelt was born on January 30, 1882. If he were alive today, about how old would he be? _____ years old

Date Time

LESSON 5•2

Math Boxes

1. Fill in the missing numbers on each number line.

a. 0.20 _____ _____ _____ _____ _____ _____ _____ _____ _____ 0.30

b. 2.6 _____ _____ _____ _____ _____ _____ 2.67 _____ _____ 2.7

2. Measure the line segment to the nearest millimeter. Record the measurement in millimeters and centimeters.

a. About _____ mm

b. About _____ cm _____ mm

c. About _____._____ cm

SRB 128 129

3. Tell whether each number sentence is true or false. Write T for true or F for false.

a. 7.89 − 3.36 = 4.53 _____

b. 12.6 − 4.8 = 3.9 + 3.9 _____

c. (4.6 + 2.9) − 3.1 < 3.7 _____

d. 0.20 > 0.68 − (0.42 + 0.11) _____

SRB 34–37 148

4. Insert > or < to make a true number sentence.

a. 4,500,999 _____ 879,662

b. 23,468,000 _____ 23,467,000

c. 568,009,352 _____ 568,010,320

d. 400,632 _____ 399,800

SRB 6

5. Divide mentally.

a. 45 / 9 = _____

b. 450 / 90 = _____

c. 1,000 / 200 = _____

d. _____ = 2,000 / 40

e. _____ = 6,300 / 700

SRB 21

Date Time

LESSON 5•3

Math Boxes

1. A number has

2 in the hundreds place,
7 in the tenths place,
6 in the hundredths place,
4 in the ones place,
5 in the tens place, and
1 in the thousandths place.

Write the number.

___ ___ ___ . ___ ___ ___

SRB 31

2. Solve mentally.

a. 3 * 40 = _____

b. 90 * 70 = _____

c. 50 * _____ = 3,000

d. _____ * 8 = 4,000

e. 80 * _____ = 56,000

SRB 17

3. Solve mentally or with a paper-and-pencil algorithm.

a.
```
    72
   450
   160
+ 1,000
```

b.
```
    15
   240
   350
+ 5,600
```

SRB 10 11

4. List all the factors of these numbers.

a. 50 _____

_____

b. 32 _____

_____

c. 60 _____

_____

SRB 7

5. If 1 centimeter on a map represents 200 miles, what do 4.5 centimeters represent? Fill in the circle next to the best answer.

○ **A.** 850 miles

○ **B.** 900 miles

○ **C.** 450 miles

○ **D.** 800 miles

SRB 43

6. a. Sara collected 30 leaves. On the way to school, she lost 2 of them. At school she and her 6 friends shared them equally. How many leaves did each person get?

_____ leaves

b. Ava and her 3 sisters shared 24 mints equally. How many mints did each sister get?

_____ mints

SRB 20

Date Time

LESSON 5•3

Estimated U.S. Distances and Driving Times

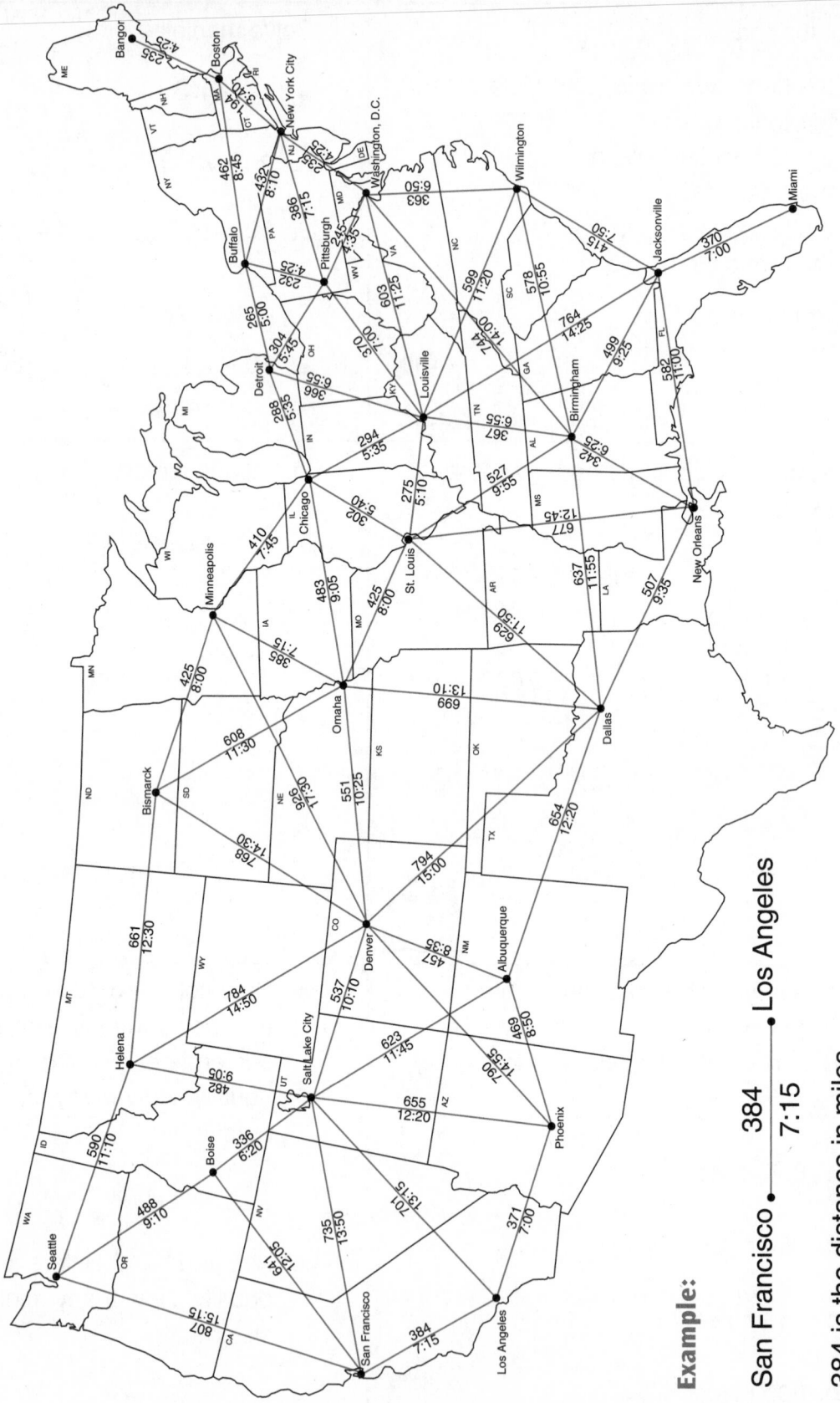

Example:

San Francisco — 384 / 7:15 — Los Angeles

384 is the distance in miles
7:15 is the driving time (hours:minutes)

Date Time

LESSON 5•3

Planning a Driving Trip

Use the map on journal page 112. Start at your hometown. Plan a driving trip that takes you to 4 other cities on the map. If your hometown is not on the map, find the nearest city on the map to your hometown. Start your driving trip from this city.

Example: Start in Chicago. Drive to St. Louis, Louisville, Birmingham, and then New Orleans.

1. Record your routes, driving distances, and driving times in the table.

From...To	Driving Distance (miles)	Driving Time (hours:minutes)	Rounded Time (hours)
	___ ___ ___	___ : ___	___
	___ ___ ___	___ : ___	___
	___ ___ ___	___ : ___	___
	___ ___ ___	___ : ___	___

2. Estimate how many *miles* you will drive in all. About _____ miles

3. Estimate how many *hours* you will drive in all. About _____ hours

4. Tell how many *days* it will take to complete the trip if you plan to drive about 8 hours each day and then stop somewhere for the night. _____ days

5. Explain how you solved Problems 2–4.

_____

_____

_____

_____

LESSON 5•4 What Do Americans Eat?

The U.S. Department of Agriculture conducts a survey to find out how much food Americans eat. A large number of people are asked to keep lists of all the foods they eat over several days.

These lists are then used to estimate how much of each food was eaten during one year. The average American eats more than 2,000 pounds of food per year. This is about $5\frac{1}{2}$ pounds of food per day.

Results show that the average American eats or drinks about the following amounts in one year:

16	pounds of apples
27	pounds of bananas
5	pounds of broccoli
10	pounds of carrots
30	pounds of cheese
255	eggs
16	pounds of fish
28	cups of yogurt
17	pounds of ice cream
350	cups of milk
22	pounds of candy

Use your answers to the Math Message question to complete these statements.

1. I will eat about _____ eggs in one year.

2. I will drink about _____ cups of milk in one year.

3. I will eat about _____ cups of yogurt in one year.

4. Based on your answers to Problems 1–3, do you think you eat like an average American? Explain why or why not.

_____

_____

Date Time

LESSON 5•4

Estimating Averages

- Estimate whether the answer will be in the tens, hundreds, thousands, or more.
- Write a number model to show how you estimated.
- Then circle the box that shows your estimate.

Example: Alice sleeps an average of 9 hours per night. How many hours does she sleep in 1 year?

Number model: 10 * 400 = 4,000

10s	100s	1,000s	10,000s	100,000s	1,000,000s

1. An average of about 23 new species of insects are discovered each day. About how many new species are discovered in one year?

   Number model: _____

10s	100s	1,000s	10,000s	100,000s	1,000,000s

2. A housefly beats its wings about 190 times per second. That's about how many times per minute?

   Number model: _____

10s	100s	1,000s	10,000s	100,000s	1,000,000s

3. A blue whale weighs about as much as 425,000 kittens. About how many kittens weigh as much as 4 blue whales?

   Number model: _____

10s	100s	1,000s	10,000s	100,000s	1,000,000s

4. An average bee can lift about 300 times its own weight. If a 170-pound person were as strong as a bee, about how many pounds could this person lift?

   Number model: _____

10s	100s	1,000s	10,000s	100,000s	1,000,000s

Date Time

LESSON 5•4

Math Boxes

1. Fill in the missing numbers on each number line.

a.

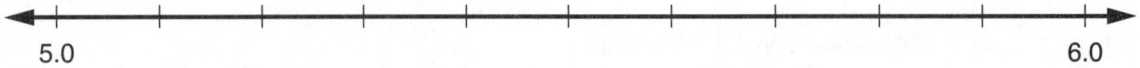

_____ _____ _____ _____ _____ _____ _____ _____ _____

b.

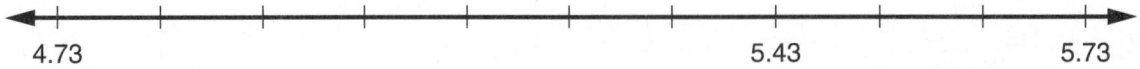

_____ _____ _____ _____ _____ _____ _____ _____

2. Measure the line segment to the nearest millimeter. Record the measurement in millimeters and centimeters.

_____

a. About _____ mm

b. About _____ cm _____ mm

c. About _____.____ cm

SRB 128 129

3. Which number sentence is true? Fill in the circle next to the best answer.

○ **A.** $(7.6 + 3.8) - 5.2 < 5.9$

○ **B.** $0.50 < 1.43 - (0.77 + 0.19)$

○ **C.** $14.46 - 12.09 = 3.46$

○ **D.** $15.8 - 11.5 = 1.6 + 2.7$

SRB 34–37 148

4. Insert > or < to make a true number sentence.

a. 9,000,000 _____ 8,999,999

b. 421,936,500 _____ 422,985,600

c. 68,004,002 _____ 68,004,100

d. 600,523 _____ 601,000

SRB 6

5. Divide mentally.

a. 18,000 / 9 = _____

b. 350 / 7 = _____

c. 3,500 / 5 = _____

d. _____ = 5,600 / 800

e. _____ = 42,000 / 60

SRB 21

LESSON 5•5

Math Boxes

1. In 2002, about 32,500,000 people living in the United States had been born in other countries. The circle graph shows where these people were born.

   a. Where were most of these people born?

   _____

   b. About what fraction of the people were born in Asia?_____

Immigration Statistics

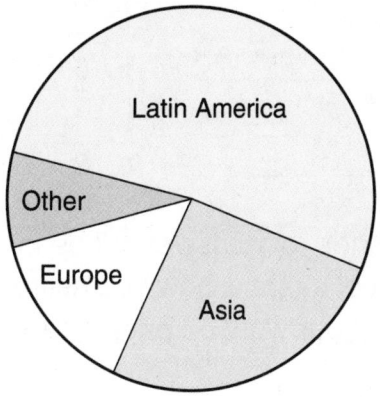

2. Estimate the sum. Write a number model to show how you estimated.

   a. 387 + 945 + 1,024

   Number model:

   _____

   b. 582 + 1,791 + 2,442

   Number model:

   _____

SRB 181

3. Complete.

   Rule: _____

in	out
4	280
80	
	490
90	6,300
300	

SRB 162–166

4. Write two hundred million, three thousand, eighty-eight using digits. Fill in the circle next to the best answer.

   ○ **A.** 2,030,088

   ○ **B.** 200,030,088

   ○ **C.** 20,003,880

   ○ **D.** 200,003,088

SRB 4

5. Look at the grid below.

   a. In which column is the circle located?

   _____

   b. In which row is the circle located?

   _____

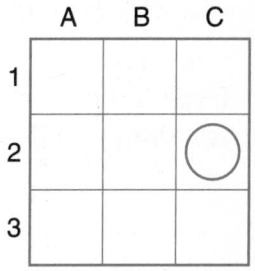

SRB 144

Date Time

LESSON 5•5

The Partial-Products Method

Multiply using the partial-products method. Show your work in the computation grid on page 119.

1. 82 * 3 = _____
2. 6 * 53 = _____
3. 574 * 5 = _____
4. 3 * 470 = _____
5. 2 * 1,523 = _____
6. 3,467 * 3 = _____

Estimate whether your answer will be in the tens, hundreds, thousands, or more. Write a number model to show how you estimated. Circle the correct box. Then calculate the answer. Show your work on page 119.

7. China has the world's longest school year at 251 days. How many school days are in 7 years?

   a. Number model: _____

10s	100s	1,000s	10,000s	100,000s	1,000,000s

   b. Calculate the answer. _____ days of school

8. People living in the United States eat about 126 pounds of fresh fruit in one year. About how many pounds of fresh fruit would a family of 6 eat in one year?

   a. Number model: _____

10s	100s	1,000s	10,000s	100,000s	1,000,000s

   b. Calculate the answer. About _____ pounds of fresh fruit

9. Explain how estimation can help you decide whether an answer to a multiplication problem makes sense.

_____

_____

_____

LESSON 5•5

The Partial-Products Method *continued*

Date Time

LESSON 5•5

My Measurement Collection for Units of Length

Use your personal references on journal page 98 to estimate the length or height of an object or distance in inches, feet, or yards. Describe the object or distance, and record your estimate in the table below. Then measure the object or distance, and record the actual measurement in the table.

SRB 128 130

Object or Distance	Estimated Length	Actual Length

Date Time

LESSON 5•6

Math Boxes

1. a. Measure the line segment to the nearest $\frac{1}{4}$ inch.

About _____ inches

b. Draw a line segment that is half as long as the one above.

c. How long is the line segment you drew? About _____ inches

SRB 128

2. Estimate the product. Write a number model to show how you estimated.

a. 48 ∗ 21

Number model:

_____

b. 98 ∗ 72

Number model:

_____

SRB 184

3. Multiply. Use the partial-products method.

_____ = 52 ∗ 43

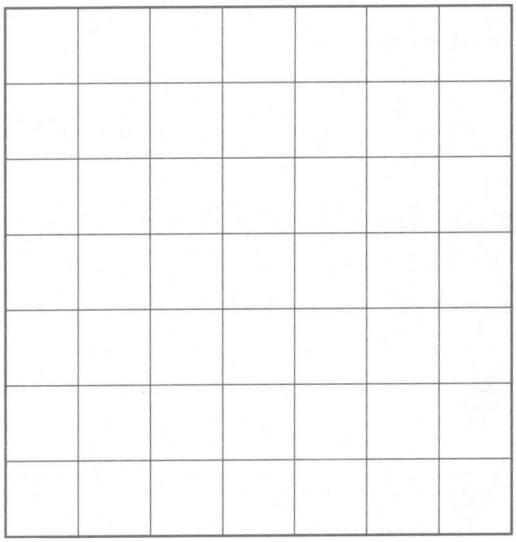

SRB 18

4. Write each number using digits.

a. three hundred forty-two thousandths

_____

b. six and twenty-five hundredths

_____

SRB 27 28

5. If you remove 7 gallons per day from a 65-gallon water tank, how many days will it take to empty the tank?

_____

SRB 175

Date Time

LESSON 5•6 Multiplication Number Stories

Follow these steps for each problem.

SRB 17 18 184

a. Decide which two numbers need to be multiplied to give the exact answer. Write the two numbers.

b. Estimate whether the answer will be in the tens, hundreds, thousands, or more. Write a number model for the estimate. Circle the box to show your estimate.

c. On the grid below, find the exact answer by multiplying the two numbers. Write the answer.

1. The average person in the United States drinks about 61 cups of soda per month. About how many cups of soda is that per year?

a. _____ * _____ (numbers that give the exact answer)

b. _____ (number model for your estimate)

c. _____ (exact answer)

10s	100s	1,000s	10,000s	100,000s	1,000,000s

2. Eighteen newborn hummingbirds weigh about 1 ounce. About how many of them does it take to make 1 pound? (1 pound = 16 ounces)

a. _____ * _____ (numbers that give the exact answer)

b. _____ (number model for your estimate)

c. _____ (exact answer)

10s	100s	1,000s	10,000s	100,000s	1,000,000s

Date Time

LESSON 5•6

Multiplication Number Stories *continued*

3. A test found that a lightbulb lasts an average of 63 days after being turned on. About how many hours is that?

a. _____ * _____ **b.** _____ **c.** _____

numbers that give the exact answer — number model for your estimate — exact answer

10s	100s	1,000s	10,000s	100,000s	1,000,000s

4. A full-grown oak tree loses about 78 gallons of water through its leaves per day. About how many gallons of water is that per year?

a. _____ * _____ **b.** _____ **c.** _____

numbers that give the exact answer — number model for your estimate — exact answer

10s	100s	1,000s	10,000s	100,000s	1,000,000s

Date Time

LESSON 5•7

Lattice Multiplication

Use the lattice method to find the products.

SRB 19

1. 3 * 56 = _____

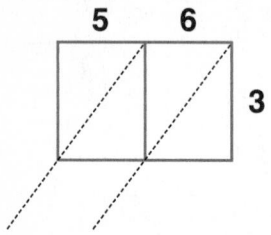

2. 8 * 26 = _____

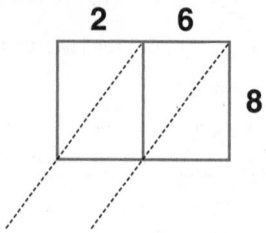

3. 7 * 74 = _____

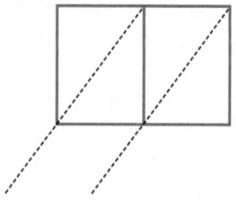

4. 6 * 315 = _____

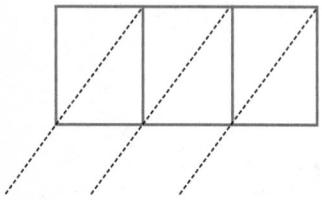

5. 9 * 284 = _____

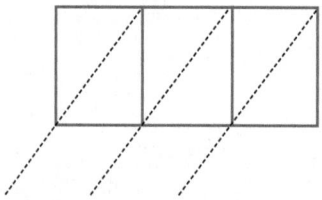

6. 47 * 63 = _____

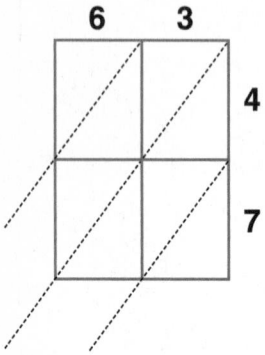

7. 26 * 26 = _____

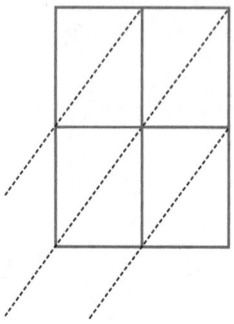

8. 16 * 473 = _____

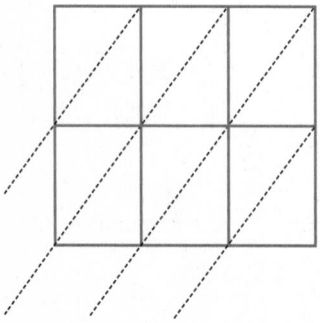

9. 46 * 805 = _____

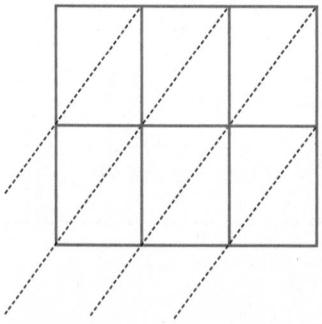

Date Time

LESSON 5•7

Math Boxes

1. This is a circle graph of what Seema does on a typical day.

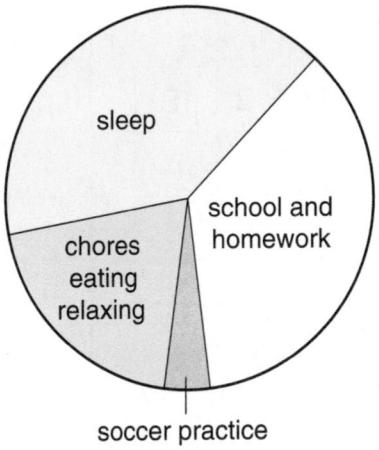

a. What does she spend the least amount of time doing?

_____

b. She spends about the same amount of time at school and doing homework as she does

_____.

c. About what fraction of the day does she spend doing chores, eating, and relaxing? _____

2. Estimate the sum. Write a number model to show how you estimated.

a. 715 + 1,904 + 688

Number model:

_____

b. 867 + 2,346 + 3,596

Number model:

_____

SRB 181

3. Complete.

Rule: _____

in	out
50	4,000
70	
	7,200
100	
45	3,600

SRB 162–166

4. Write each number using digits.

a. twenty-six million, nineteen thousand, eighteen

_____

b. three hundred fifty-two million, eight hundred thousand, two hundred

_____

SRB 4

5. Look at the grid below.

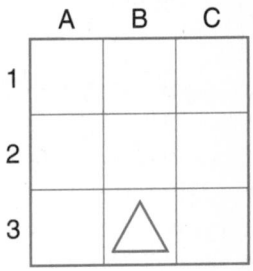

a. In which column is the triangle located?

_____

b. In which row is the triangle located?

_____

SRB 144

Date Time

LESSON 5•8

Reading and Writing Big Numbers

1. Each row in the place-value chart shows a number. Use words to write the name for each number below the chart.

SRB 4

	Billions				Millions				Thousands				Ones		
	100B	10B	1B	,	100M	10M	1M	,	100Th	10Th	1Th	,	100	10	1
a.			7	,	4	0	0	,	0	6	5	,	2	0	0
b.						5	1	,	8	0	0	,	0	0	0
c.		2	3	,	0	0	0	,	0	0	5	,	1	4	0
d.	1	2	3	,	4	5	6	,	7	8	9	,	0	1	2

a. 7 billion, 400 million, 65 thousand, 200

b. _____

c. _____

d. _____

2. Use digits to write these numbers in the place-value chart below.

a. 400 thousand, 500

b. 208 million, 350 thousand, 600

c. 16 billion, 210 million, 48 thousand, 715

d. 1 billion, 1 million, 1 thousand, 1

	Billions				Millions				Thousands				Ones		
	100B	10B	1B	,	100M	10M	1M	,	100Th	10Th	1Th	,	100	10	1
a.															
b.															
c.															
d.															

Date Time

LESSON 5•8 How Much Are a Million and a Billion?

1. How many dots are on the 50-by-40 array page? _____ dots

2. How many dots would be on

   a. 5 pages? _____ dots

   b. 50 pages? _____ dots

   c. 500 pages? _____ dots

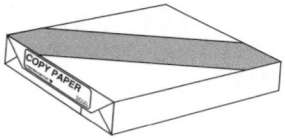

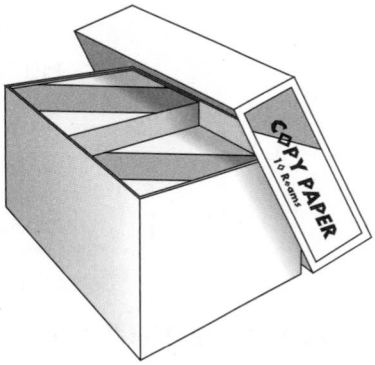

3. Each package of paper, or ream, contains 500 sheets. How many dots would be on the paper in

   a. 1 ream? (*Hint:* Look at Problem 2.) _____ dots

   b. 10 reams? (1 carton) _____ dots

   c. 100 reams? (10 cartons) _____ dots

   d. 1,000 reams? (100 cartons) _____ dots

4. Use digits to write these numbers in the place-value chart below.

   a. 999 thousand b. 1,000 thousand c. 999 million d. 1,000 million

	Billions				Millions				Thousands				Ones		
	100B	10B	1B	,	100M	10M	1M	,	100Th	10Th	1Th	,	100	10	1
a.															
b.															
c.															
d.															

Date Time

LESSON 5•8

Internet Users

The table below shows the estimated number of people who used the Internet in several different countries in 2004.

Internet Users in 2004

Country	Users
France	25,470,000
Greece	2,710,000
Hungary	2,940,000
Italy	25,530,000
Poland	10,400,000
Spain	13,440,000

Source: The World Factbook

1. Which of these countries had the most Internet users? _____

2. Which of these countries had the fewest Internet users? _____

3. About how many more Internet users were there in France than in Spain?

   Number model: _____

   Answer: _____

4. Write true or false.

   a. There were more Internet users in Greece, Poland, and Hungary combined than in Spain. _____

   b. There were about 6 times as many Internet users in France as there were in Hungary. _____

5. Why do you think the Internet usage data is rounded to the nearest 10,000 instead of an actual count?

   _____

   _____

6. The United States had about 62 times as many Internet users as Hungary. About how many Internet users were there in the United States?

   Number model: _____

   Answer: _____

LESSON 5•8

Math Boxes

1. a. Measure the line segment to the nearest $\frac{1}{4}$ inch.

R S

About _____ inches

b. Draw a line segment that is half as long as the one above.

c. How long is the line segment you drew? About _____ inches

SRB 128

2. Estimate the product. Write a number model to show how you estimated.

a. 41 * 83

Number model:

_____

b. 75 * 32

Number model:

_____

SRB 184

3. Multiply. Use the partial-products method.

_____ = 46 * 98

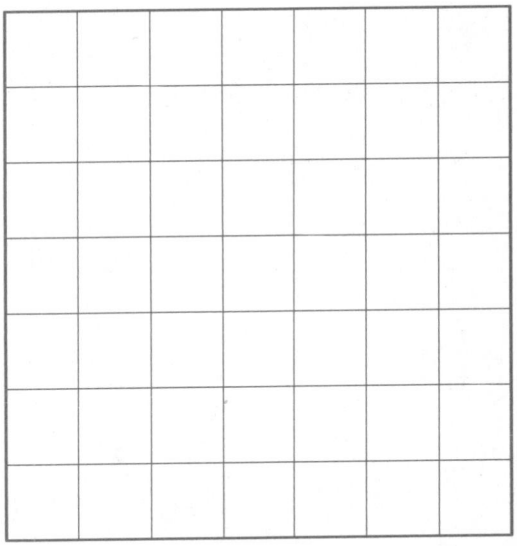

SRB 18

4. Write each number using digits.

a. seven hundred six thousandths

_____

b. three and four hundredths

_____

SRB 27 28

5. You have 75 apples. You need 8 apples to make a pie. How many pies can you bake?

○ **A.** 9

○ **B.** 8

○ **C.** 7

○ **D.** 6

SRB 175

LESSON 5•9

Place Value and Powers of 10

Millions	Hundred Thousands	Ten Thousands	Thousands	Hundreds	Tens	Ones
1,000,000			1,000	100		1
10 [100,000s]			10 [100s]	10 [10s]		10 [tenths]
		$10 * 10 * 10 * 10$		$10 * 10$		
	10^5		10^3	10^2		10^0

Fill in this place-value chart as follows:

1. Write standard numbers in Row 1.
2. In Row 2, write the value of each place to show that it is 10 times the value of the place to its right.
3. In Row 3, write the place values as products of 10s.
4. In Row 4, show the values as powers of 10. Use exponents. The exponent shows how many times 10 is used as a factor. It also shows how many zeros are in the standard number.

LESSON 5•9

Math Boxes

1. Estimate the sum. Write a number model to show how you estimated.

a. 1,254 + 8,902 + 2,877

Number model:

_____

b. 12,645 + 7,302 + 15,297

Number model:

_____

SRB 181

2. Which number sentence is true? Fill in the circle next to the best answer.

○ **A.** 5,800,000 = 58 million

○ **B.** 62 million > 3,100,000,000

○ **C.** $10^3 = 1{,}000$

○ **D.** $100{,}000 = 10^2$

SRB 5 6

3. Draw lines to match each word to the correct pair or pairs of line segments.

perpendicular

parallel

intersecting

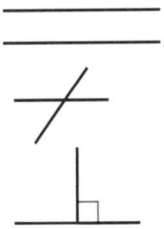

SRB 94 95

4. Complete.

Rule: _____

in	out
1.29	1.36
6.47	
	5.17
12.66	
7.93	8.00

SRB 162–166

5. Multiply. Use a paper-and-pencil algorithm.

9 ∗ 258 = _____

SRB 18 19

6. Which of the angles below has a measure of about 90 degrees? Circle it.

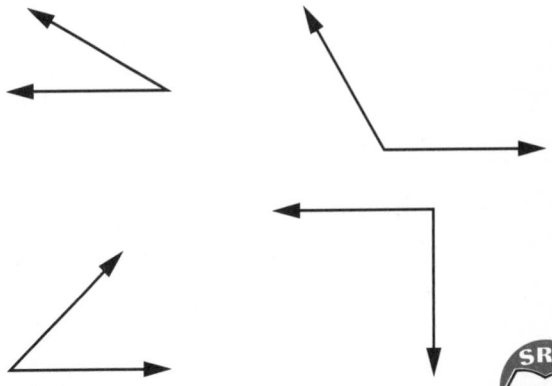

SRB 92 93

Date Time

LESSON 5•10

Evaluating Large Numbers

1. Round the attendance figures in the table to the nearest hundred-thousand.

2004 Major League Baseball Home Game Attendance for 10 Teams

Team	Total Home Game Attendance	Attendance Rounded to the Nearest 100,000
Atlanta Braves	2,603,484	
Baltimore Orioles	2,682,917	
Boston Red Sox	2,650,063	
Cleveland Indians	2,616,940	
Colorado Rockies	2,737,838	
Los Angeles Dodgers	3,131,255	
New York Yankees	3,465,807	
Seattle Mariners	3,540,658	
St. Louis Cardinals	3,011,216	
Texas Rangers	2,352,397	

Source: Information Please—Baseball Attendance

2. How do you think attendance figures for major league baseball games are obtained?

_____

_____

_____

3. Do you think *exactly* 2,737,838 people were at the home games played by the Colorado Rockies? Explain your answer.

_____

_____

_____

4. You rounded the figures in the table above to the nearest hundred-thousand. Which teams have the same attendance figures based on these rounded numbers?

_____

_____

Date Time

LESSON 5•10

Math Boxes

1. a. Measure the line segment to the nearest $\frac{1}{4}$ inch.

T G

About _____ inches

b. Draw a line segment that is half as long as the one above.

c. How long is the line segment you drew? About _____ inches

SRB 128

2. Estimate the product. Write a number model to show how you estimated.

a. 37 ∗ 91

Number model:

_____

b. 53 ∗ 17

Number model:

_____

SRB 184

3. Multiply. Use the partial-products method.

_____ = 43 ∗ 89

SRB 18

4. Write fourteen and three-tenths using digits. Fill in the circle next to the best answer.

○ **A.** 14.03

○ **B.** 14.003

○ **C.** 14.310

○ **D.** 14.3

SRB 27 28

5. There are 60 trading cards. Each student gets 5 cards. How many students get trading cards?

_____ students

SRB 175

Date Time

LESSON 5•11 Traveling to Europe

It is time to leave Africa. Your destination is Region 2—the continent of Europe. You and your classmates will fly from Cairo, Egypt to Budapest, Hungary. Before exploring Hungary, you will collect information about the countries in Region 2. You may even decide to visit another country in Europe after your stay in Budapest.

Use the World Tour section of your *Student Reference Book* to answer the questions.

1. Which country in Region 2 has

a. the largest population? _____
country population

b. the smallest population? _____
country population

c. the largest area? _____
country area

d. the smallest area? _____
country area

Use the Climate and Elevation of Capital Cities table on page 297.

2. From December to February, which capital in Region 2 has

a. the warmest weather? _____
capital country temperatures

b. the coolest weather? _____
capital country temperatures

c. the greatest amount of rain? _____
capital country inches rainfall

d. the least amount of rain? _____
capital country inches rainfall

Use the Population Data table on page 301.

3. Which country in Region 2 has

a. the greatest percent of population ages 0–14? _____
country percent

b. the smallest percent of population ages 0–14? _____
country percent

LESSON 5•11

Math Boxes

1. Estimate the sum. Write a number model to show how you estimated.

   a. 799 + 11,304 + 48,609

   Number model:

   _____

   b. 4,382 + 6,911 + 7,035

   Number model:

   _____

SRB 181

2. Write <, >, or = to make each number sentence true.

   a. 356,789 _____ 354,999

   b. 670,000 _____ 67,000,000

   c. 62 million _____ 9,700,000

   d. 105,000,000 _____ 15,500,000

   e. 10^4 _____ 1,000

SRB 5 6

3. a. Draw a pair of parallel line segments.

   b. Draw a pair of intersecting line segments.

SRB 94 95

4. Complete.

   Rule: _____

in	out
6.46	6.58
3.08	
	11.34
25.25	
	100.1
63.09	63.21

SRB 162–166

5. Multiply. Use a paper-and-pencil algorithm.

   7 ∗ 208 = _____

SRB 18 19

6. Which of the angles below has a measure less than 90 degrees? Circle it.

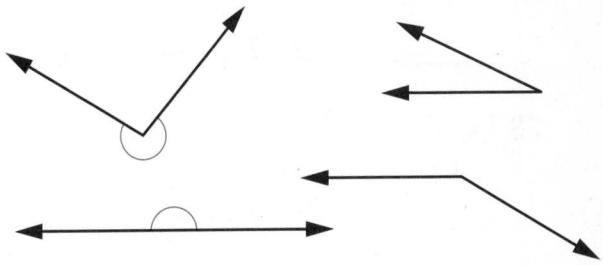

SRB 93

Date Time

LESSON 5•12

Math Boxes

1. There are 240 chairs to set up for the concert. Each row has 40 chairs in it. How many rows are there?

_____ rows

SRB 21

2. The senior class at Rees High School raised $1,895 for five charities in the community. The money will be shared equally. How much money will each charity receive?

_____

SRB 22 23

3. Look at the grid below.

a. In which column is the star located?

_____

b. In which row is the star located?

_____

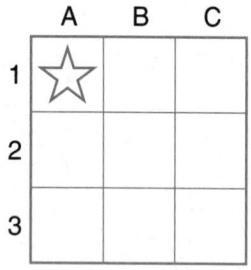

SRB 144

4. Draw a right angle with vertex *K*.

SRB 93

5. Divide.

a. 72 / 9 = _____

b. 720 / 90 = _____

c. 7,200 / 900 = _____

d. _____ = 42 / 7

e. _____ = 420 / 7

f. _____ = 4,200 / 7

g. 28 / 4 = _____

h. 28,000 / 40 = _____

i. 28,000 / 400 = _____

j. _____ = 24 / 6

k. _____ = 2,400 / 60

l. _____ = 24,000 / 6

SRB 20 21

LESSON 6•1

Math Boxes

1. Measure each line segment to the nearest millimeter.

   a. R ——— S

   About _____ cm _____ mm

   b. C ——— S

   About _____ cm _____ mm

SRB 128

2. Round 409,381,886 to the nearest

   a. hundred.

   _____

   b. ten thousand.

   _____

   c. ten million.

   _____

   d. hundred million.

   _____

SRB 182 183

3. Multiply. Use a paper-and-pencil method.

   _____ = 86 ∗ 29

SRB 18 19

4. Complete.

   a. 10^2 = _____

   b. $10^{\square}$ = 10 ∗ 10 ∗ 10 ∗ 10

   c. 1,000 = $10^{\square}$

   d. 10 to the ninth power =

   _____

SRB 5

5. Circle $\frac{5}{6}$ of the squares.

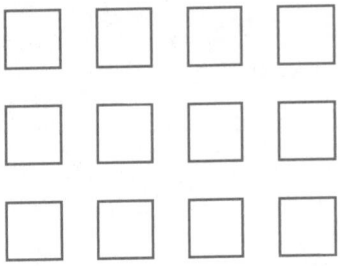

SRB 44

Date Time

LESSON 6•1

Multiplication/Division Number Stories

Fill in each Multiplication/Division Diagram. Then write a number model. Be sure to include a unit with your answer.

1. The profit from the Maple Street lemonade stand was $36. Four girls will share this amount equally. What will each girl's share be?

girls	dollars per girl	dollars in all

Number model: _____

Answer: _____

2. Sheila has 57 pictures to put in her photo album. She can put 6 pictures on each page. How many pages will be completely filled up when she is finished?

pages	pictures per page	pictures in all

Number model: _____

Answer: _____

3. Reuben walks a total of 35 blocks going to and from school each week. He always walks the same route. How many blocks does he walk each day?

days	blocks per day	blocks in all

Number model: _____

Answer: _____

LESSON 6•1

Multiplication/Division Number Stories *cont.*

4. Hassan is helping his teacher put 8 centimeter cubes into each paper cup for a math project. How many paper cups can he fill if there are 79 cubes?

paper cups	cubes per paper cup	cubes in all

Number model: _____

Answer: _____

Try This

5. Mr. Henning's fourth-grade class is planning a field trip to see a play. The bus will cost $100, and the tickets will cost $125. The 25 students will share the total cost equally. How much will each student pay for the field trip?

_____	_____ per _____	_____ in all

Number model: _____

Answer: _____

6. Martina and 3 friends sold 636 boxes of cookies for their club. Each girl sold the same number of boxes. How many boxes of cookies did each girl sell?

_____	_____ per _____	_____ in all

Number model: _____

Answer: _____

Date Time

LESSON 6•1

Extended Multiplication Facts

SRB 17

1. 9 * 5 = _____

9 * 50 = _____

90 * 5 = _____

90 * 50 = _____

900 * 5 = _____

90 * 500 = _____

2. 8 * 7 = _____

8 * 70 = _____

80 * 7 = _____

80 * 70 = _____

800 * 7 = _____

80 * 700 = _____

3. 4 * 9 = _____

4 * 90 = _____

40 * 9 = _____

40 * 90 = _____

400 * 9 = _____

40 * 900 = _____

4. 6 * _____ = 18

60 * _____ = 180

60 * _____ = 1,800

_____ * 60 = 180

_____ * 600 = 1,800

30 * _____ = 18,000

5. _____ * 8 = 48

_____ * 80 = 480

_____ * 80 = 4,800

60 * _____ = 480

6 * _____ = 4,800

6 * _____ = 48,000

6. 8 * _____ = 24

8 * _____ = 2,400

80 * _____ = 2,400

_____ * 30 = 240

_____ * 3 = 240

_____ * 300 = 240,000

LESSON 6•2

Math Boxes

1. There are 32 students in the class. A yearbook page can show 8 student photos. How many pages are needed to include all the student photos?

pages	photos per page	photos in all

Number model: _____

Answer: _____ pages

2. Solve each open sentence.

a. $24 = a * (5 + 1)$ $a =$ _____

b. $54 / 6 = 81 / b$ $b =$ _____

c. $(c + 4) / 3 = 7$ $c =$ _____

d. $m - 3.87 = 7.49$ $m =$ _____

e. $0.98 + 4.83 = f + 4.35$ $f =$ _____

SRB 148

3. Use a paper-and-pencil algorithm to add or subtract.

a. $\begin{array}{r} 0.85 \\ +\ 0.53 \\ \hline \end{array}$

b. $\begin{array}{r} 0.64 \\ +\ 1.73 \\ \hline \end{array}$

c. $\begin{array}{r} 12.38 \\ -\ 1.09 \\ \hline \end{array}$

d. $\begin{array}{r} 3.05 \\ -\ 0.67 \\ \hline \end{array}$

SRB 34–37

4. Complete.

a. 670 cm = _____ m

b. 4,800 cm = _____ m

c. 916 cm = _____ m _____ cm

d. 18 m = _____ cm

SRB 129

5. Name a fraction equivalent to $\frac{1}{2}$. Circle the best answer.

A. $\frac{3}{4}$

B. $\frac{8}{9}$

C. $\frac{5}{10}$

D. $\frac{3}{5}$

SRB 51

Date Time

LESSON 6•2

Solving Division Problems

For Problems 1–6, fill in the multiples-of-10 list if it is helpful. If you prefer to solve the division problems in another way, show your work.

1. José's class baked 64 cookies for the school bake sale. Students put 4 cookies in each bag. How many bags of 4 cookies did they make?

10 [4s] = _____

20 [4s] = _____

30 [4s] = _____

40 [4s] = _____

50 [4s] = _____

Number model: _____

Answer: _____ bags

---

2. The community center bought 276 cans of soda for a picnic. How many 6-packs is that?

10 [6s] = _____

20 [6s] = _____

30 [6s] = _____

40 [6s] = _____

50 [6s] = _____

Number model: _____

Answer: _____ 6-packs

---

3. Each lunch table at Johnson Elementary School seats 5 people. How many tables are needed to seat 191 people?

10 [5s] = _____

20 [5s] = _____

30 [5s] = _____

40 [5s] = _____

50 [5s] = _____

Number model: _____

Answer: _____ tables

Date Time

LESSON 6•2

Solving Division Problems *continued*

4. The preschool held a tricycle parade. Trent counted 135 wheels. How many tricycles is that?

10 [3s] = _____

20 [3s] = _____

30 [3s] = _____

40 [3s] = _____

50 [3s] = _____

Number model: _____

Answer: _____ tricycles

---

5. How many 8s are there in 248?

10 [8s] = _____

20 [8s] = _____

30 [8s] = _____

40 [8s] = _____

50 [8s] = _____

Number model: _____

Answer: _____

---

6. How many 7s are in 265?

10 [7s] = _____

20 [7s] = _____

30 [7s] = _____

40 [7s] = _____

50 [7s] = _____

Number model: _____

Answer: _____

Date Time

LESSON 6•3

Partial-Quotients Division Algorithm

SRB 22 23

These notations for division are equivalent:

$6\overline{)134}$ $134 \div 6$ $134 / 6$ $\frac{134}{6}$

1. There are 6 pencils in each pack. How many packs can be made from 96 pencils?

   Number model: _____

   Answer: _____ packs

   How many pencils are left over? ____ pencils

2. Phil has $79 to purchase books. Books cost $7 each. How many books can Phil buy?

   Number model: _____

   Answer: _____ books

   How many dollars are left over? ____ dollars

3. There are 184 plants to be put into pots. Each pot can hold 8 plants. How many pots are needed?

   Number model: _____

   Answer: _____ pots

   How many plants are left over? ____ plants

4. The principal shared 395 cookies equally among 9 classes. How many cookies did each class receive?

   Number model: _____

   Answer: _____ cookies

   How many cookies were left over? ____ cookies

Date Time

LESSON 6•3

Partial-Quotients Division Algorithm *cont.*

Divide.

5. $3\overline{)87}$

Answer: _____

6. 331 ÷ 7

Answer: _____

Try This

7. Twelve shirts fit into a box. There are 372 shirts to be put into boxes. How many boxes are needed?

Number model: _____

Answer: _____ boxes

How many shirts are left over? ____ shirts

8. There are _____ players in the league. (Write a number greater than 100.)

There are ____ players on each team. (Write a number between 3 and 9.)

How many teams can be made?

Number model: _____

Answer: _____ teams

How many players are left over? ____ players

Date Time

LESSON 6•3

Place Value in Decimals

SRB 30 31

1. Write these numbers in order from smallest to largest.

   1.26 0.58 1.09 1.091 0.35

   _____ smallest

   _____

   _____

   _____

   _____ largest

2. A number has

   6 in the tenths place,
   4 in the ones place,
   5 in the hundredths place, and
   9 in the tens place.

   Write the number.

   ____ ____ . ____ ____

3. Write the smallest number you can make with the following digits:

   3 6 4 7 2

   _____

4. What is the value of the digit 4 in the numerals below?

   a. 37.48 _____

   b. 49.08 _____

   c. 0.942 _____

   d. 1.664 _____

5. Write each number using digits.

   a. four and seventy-two hundredths

   _____

   b. nine hundred thirty-five thousandths

   _____

6. I am a four-digit number less than 10.

   - The digit in the tenths place is the result of dividing 36 by 4.
   - The digit in the hundredths place is the result of dividing 42 by 7.
   - The digit in the ones place is the result of dividing 72 by 8.
   - The digit in the thousandths place is the result of dividing 35 by 5.

   What number am I?

   ____ . ____ ____ ____

LESSON 6•3

Math Boxes

1. Measure each line segment to the nearest millimeter.

a. _____
P *S*

About _____ cm _____ mm

b. _____
A *B*

About _____ cm _____ mm

SRB 128

2. Round 5,906,245 to the nearest

a. million.

_____

b. ten thousand.

_____

c. thousand.

_____

d. hundred.

_____

SRB 182 183

3. Multiply. Use a paper-and-pencil method.

_____ = 58 ∗ 52

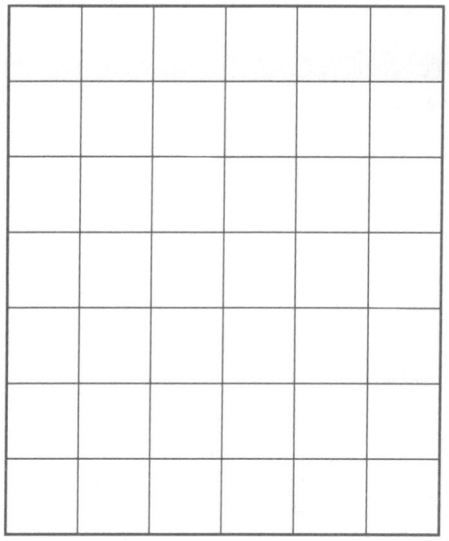

SRB 18 19

4. Complete.

a. 10^4 = _____

b. $10^{\square} = 10 * 10 * 10 * 10 * 10$

c. $100 = 10^{\square}$

d. 10 to the seventh power =

_____

SRB 5

5. Circle $\frac{1}{2}$ of the squares.

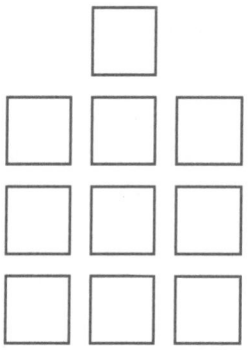

SRB 44

Date Time

LESSON 6•4

Interpreting Remainders

For each number story:

- Draw a picture.
- Write a number model.
- Use a division algorithm to solve the problem.
- Decide what to do about the remainder.

1. Jackson is buying balloons for a party. Balloons cost $6 per bunch. How many bunches can he buy with $75?

Picture:

Number model: _____

Answer: _____ bunches

What did you do about the remainder? Circle the answer.

A. Ignored it

B. Reported it as a fraction or decimal

C. Rounded the answer up

Why? _____

_____

_____

2. Rosa is buying boxes to hold all 128 of her CDs. Each box holds 5 CDs. How many boxes are needed to store all of her CDs?

Picture:

Number model: _____

Answer: _____ boxes

What did you do about the remainder? Circle the answer.

A. Ignored it

B. Reported it as a fraction or decimal

C. Rounded the answer up

Why? _____

_____

_____

LESSON 6•4 Interpreting Remainders *continued*

3. Lateefah won 188 candy bars in a raffle. She decided to share them equally with 7 of her classmates and herself. How many candy bars did each person receive?

Picture:

Number model: _____

Answer: _____ candy bars

What did you do about the remainder? Circle the answer.

A. Ignored it

B. Reported it as a fraction or decimal

C. Rounded the answer up

Why? _____

_____

_____

_____

Try This

Write each answer as a mixed number by rewriting the remainder as a fraction.

4. 2)27 = 13 R1 _____

5. 10)883 = 88 R3 _____

6. 16)252 = 15 R12 _____

Write each answer as a decimal.

7. 39 ÷ 2 = 19 R1 _____

8. 183 ÷ 12 = 15 R3 _____

9. 2,067 ÷ 5 = 413 R2 _____

Date Time

LESSON 6•4

Math Boxes

1. Joe ordered 72 plants for his patio garden. Each pot holds 4 plants. How many pots are needed to hold all of the plants?

pots	plants per pot	plants in all

Number model: _____

Answer: _____

2. Solve each open sentence.

a. $(6 + 9) + (3 * A) = 30$ $A =$ _____

b. $24 \div 8 = 21 \div B$ $B =$ _____

c. $72 = (2 * C) * 9$ $C =$ _____

d. $6.2 + 0.79 = D$ $D =$ _____

e. $8.91 - E = 2.72$ $E =$ _____

SRB 35–37 148

3. Use a paper-and-pencil algorithm to add or subtract.

a. $\begin{array}{r} 0.37 \\ +\ 0.26 \\ \hline \end{array}$

b. $\begin{array}{r} 2.9 \\ +\ 5.01 \\ \hline \end{array}$

c. $\begin{array}{r} 6.79 \\ -\ 6.55 \\ \hline \end{array}$

d. $\begin{array}{r} 7.80 \\ -\ 3.65 \\ \hline \end{array}$

SRB 34–37

4. How many centimeters are in 12 meters? Circle the best answer.

A. 0.12

B. 1.2

C. 120

D. 1,200

SRB 129

5. Circle the fractions equivalent to $\frac{1}{2}$.

$\frac{8}{16}$ $\frac{5}{6}$ $\frac{6}{12}$

$\frac{2}{3}$ $\frac{12}{24}$ $\frac{8}{15}$

SRB 51

Date Time

LESSON 6•5

Math Boxes

1. Insert parentheses to make each number sentence true.

   a. $15 + 5 * 6 = 120$

   b. $7 + 9 * 2 = 25$

   c. $77 = 1 + 6 * 6 + 5$

SRB 150

2. Draw a line segment that is 2 inches long. Mark and label the following inch measurements on the line segment:

   $\frac{1}{2}$, $\frac{3}{4}$, 1, $1\frac{1}{2}$, and 2

SRB 128

3. The Sports Boosters raised $908 at their annual chili supper. Four athletic teams will share the money equally.

   How much money will each team receive?

   Number model: _____

   Answer: _____

SRB 22 23

4. Multiply with a paper-and-pencil algorithm.

   $66 * 62 =$ _____

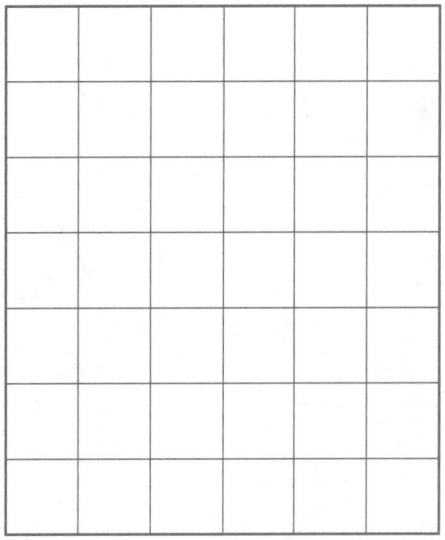

SRB 18 19

5. Complete.

   a. 9 m = _____ cm

   b. 1,500 cm = _____ m

   c. 350 cm = _____ m

   d. 458 cm = _____ m _____ cm

   e. 3.2 m = _____ cm

SRB 129

6. a. Shade $\frac{1}{2}$ of the square.

   b. Shade $\frac{2}{3}$ of the square.

SRB 44

Date Time

LESSON 6•5 Making a Full-Circle Protractor

There are 360 marks around the circle. They divide the edge of the circle into 360 small spaces. Twelve of the marks are longer than the rest. They are in the same positions as the 12 numbers around a clock face. Your teacher will tell you how to label the 12 large marks on the circle.

SRB 92

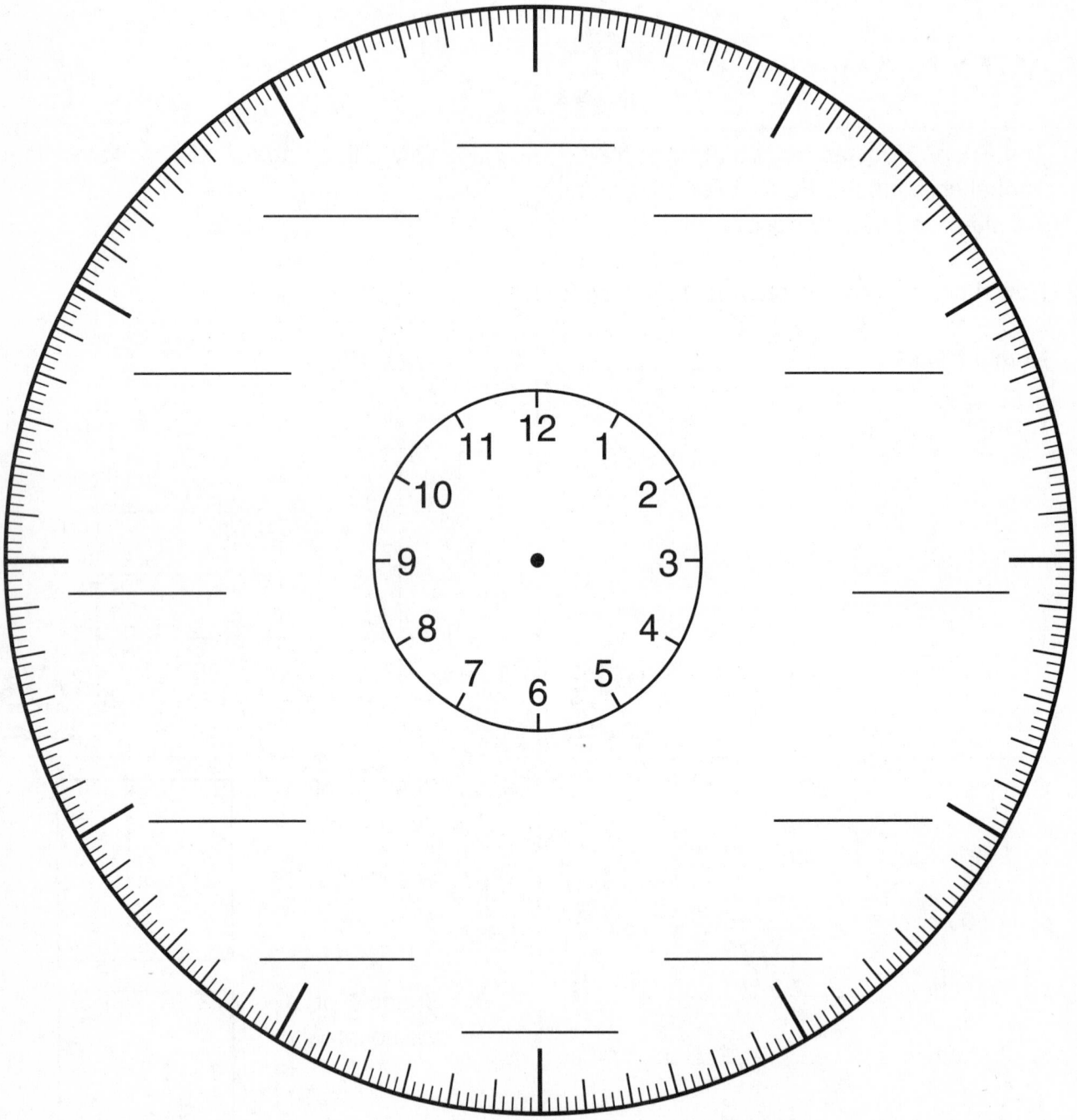

Date Time

LESSON 6•5 Clock Angles

Use the clock below and the full-circle protractor on journal page 152 to help you answer the questions.

SRB 92 141

1. How many minutes and how many degrees does the *minute hand* move

   a. from 3:00 to 4:00? _____ minutes _____ °

   b. from 7:00 to 7:45? _____ minutes _____ °

   c. from 8:15 to 8:45? _____ minutes _____ °

   d. from 6:30 to 6:50? _____ minutes _____ °

   e. from 5:15 to 5:30? _____ minutes _____ °

   f. from 1:00 to 1:10? _____ minutes _____ °

   g. from 12:00 to 12:05? _____ minutes _____ °

   h. from 5:00 to 5:01? _____ minutes _____ °

Try This

2. How many degrees does the *hour hand* move

   a. in 1 hour? _____ °

   b. in $\frac{1}{2}$ hour? _____ °

   c. in 10 minutes? _____ °

3. Explain how you solved Problem 2c.

_____

_____

_____

_____

Date Time

LESSON 6•5

Population Bar Graph

The table below shows the percent of the population (number of people out of 100) who are 14 years old or younger in the Region 2 countries.

Country	Percent of Population Ages 0–14
France	19
Greece	15
Hungary	16
Iceland	23
Italy	14
Netherlands	18
Norway	20
Poland	18
Spain	15
United Kingdom	19

SRB 76 301

1. Make a bar graph to display the information given in the table above.

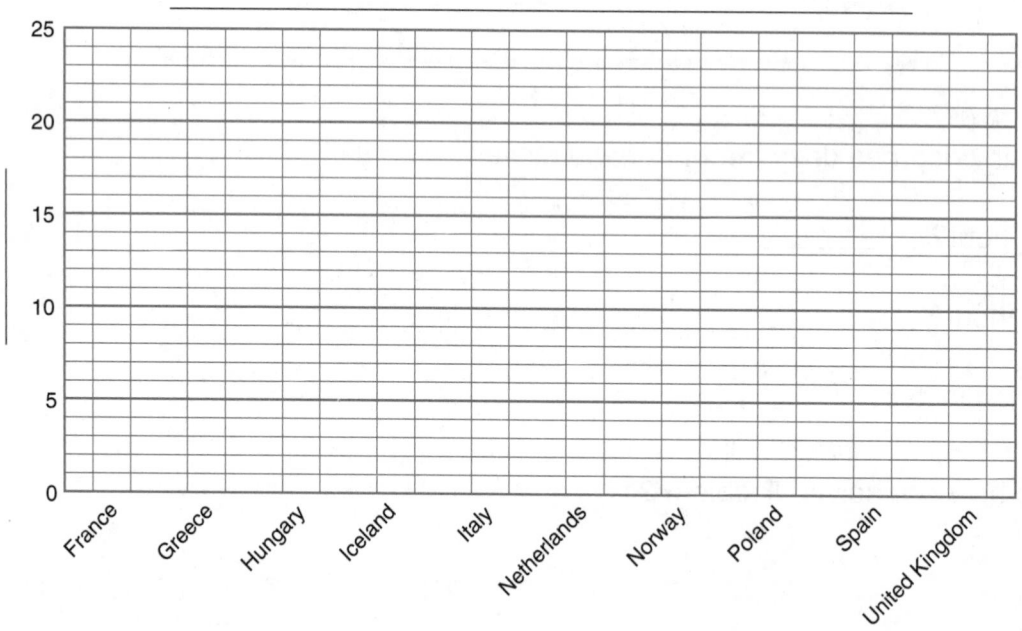

2. Why might it be important to know what percent of the population of a country is 0 through 14 years of age?

Date Time

LESSON 6•6

Measuring Angles

Use your full-circle protractor to measure each angle.

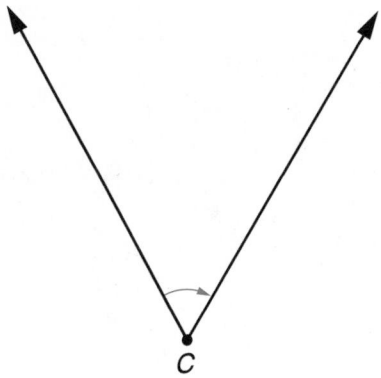

1. $\angle C$ measures _____ °.

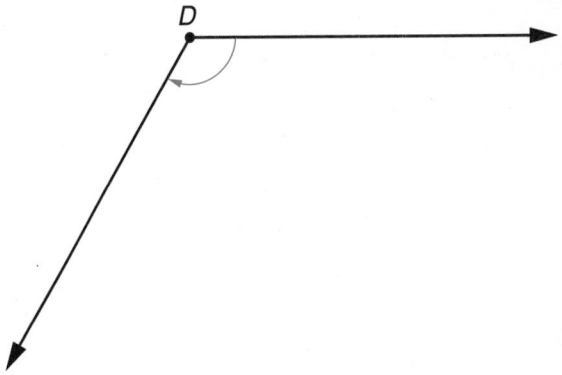

2. $\angle D$ measures _____ °.

Try This

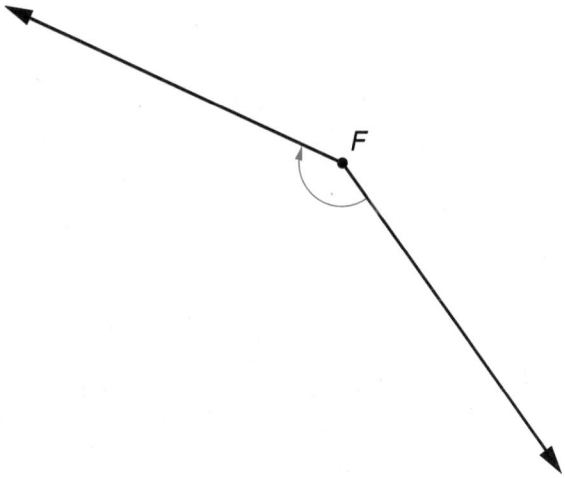

3. $\angle F$ measures _____ °.

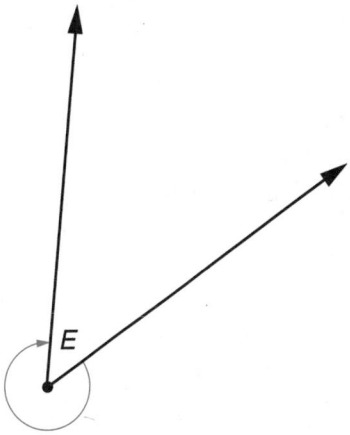

4. $\angle E$ measures _____ °.

5. Without using your full-circle protractor, give the measure of the reflex angle in Problem 3 (the part not marked by the blue arrow). Explain your answer.

_____

_____

_____

LESSON 6•6

Math Boxes

1. Ms. Kawasaki's fourth grade class made a circle graph to show students' favorite days of the week.

   a. Which day of the week is the least favorite in Ms. Kawasaki's classroom?

   _____

   b. About what fraction of the students prefer Saturday?

   _____

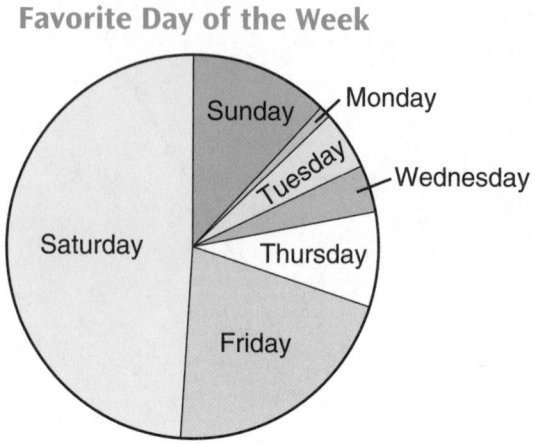

2. Juan talked on the phone an average of 34 minutes per week for 1 whole year. About how many minutes did Juan spend on the phone in 1 year?

   Number model: _____

   Answer: _____ minutes

3. Divide with a paper-and-pencil algorithm. Write the remainder as a fraction.

   883 / 7 = _____

4. Write <, >, or = to make each number sentence true.

   a. 420,000,000 _____ four hundred twenty million

   b. 65,000,000 _____ 92,000,000

   c. four hundred thousand _____ 10^4

   d. 10^2 _____ 1,000

5. For this spinner, what color would you be *most likely* to land on?

   _____

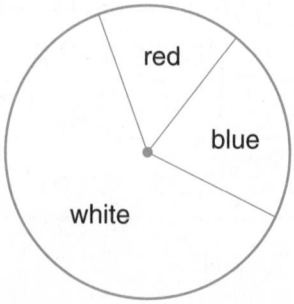

Date Time

LESSON 6•7 Drawing and Measuring Angles

Math Message

Use a straightedge to draw the following angles. Do **not** use a protractor.

$\angle A$: any angle less than 90°

$\angle B$: any angle more than 90° and less than 180°

$\angle C$: any angle more than 180°

$\angle A$ is called an **acute angle.**

$\angle B$ is called an **obtuse angle.**

$\angle C$ is called a **reflex angle.**

Measuring Angles with a Protractor

Write whether the angle is *acute* or *obtuse*. Then measure it as accurately as you can.

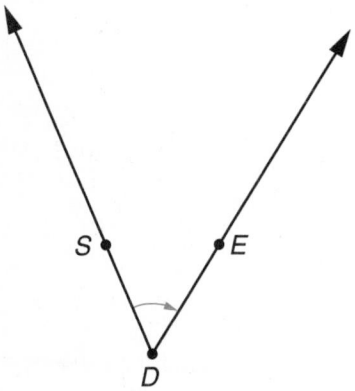

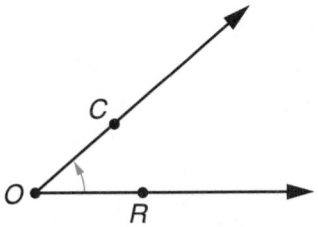

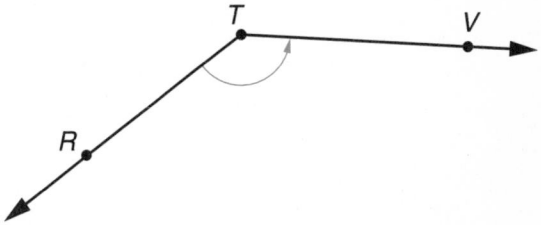

$\angle SDE$ is _____.

$\angle SDE$ is about _____ °.

$\angle COR$ is _____.

$\angle COR$ is about _____ °.

$\angle RTV$ is _____.

$\angle RTV$ is about _____ °.

Date Time

LESSON 6•7

Drawing Angles

1. Draw a 35° angle, using line segment *GH* as one of its sides.

2. Draw a 150° angle, using ray *CD* as one of its sides.

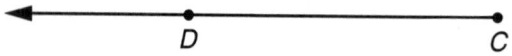

3. Draw a 60° angle, using ray *EF* as one of its sides.

4. Draw a 15° angle, using ray *AB* as one of its sides.

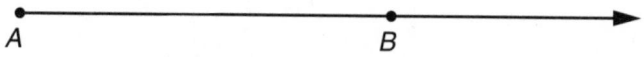

Try This

5. Draw a 330° angle, using ray *IJ* as one of its sides.

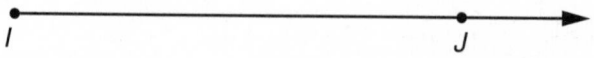

LESSON 6•7

Math Boxes

1. Insert parentheses to make each number sentence true.

a. $12 = 15 - 2 + 1$

b. $66 - 16 * 4 = 200$

c. $49 = 4 + 3 * 42 / 6$

SRB 150

2. Draw a line segment that is 2 inches long. Mark and label the following inch measurements on the line segment:

$\frac{1}{4}$, $\frac{3}{4}$, 1, $1\frac{1}{4}$ and $1\frac{1}{2}$

SRB 128

3. Six classrooms collected newspapers for one week. If they collected a total of 582 newspapers by the end of the week, on average about how many newspapers did each class collect?

Number model: _____

Answer: _____ newspapers

SRB 22 23

4. Multiply with a paper-and-pencil algorithm.

$67 * 34 =$ _____

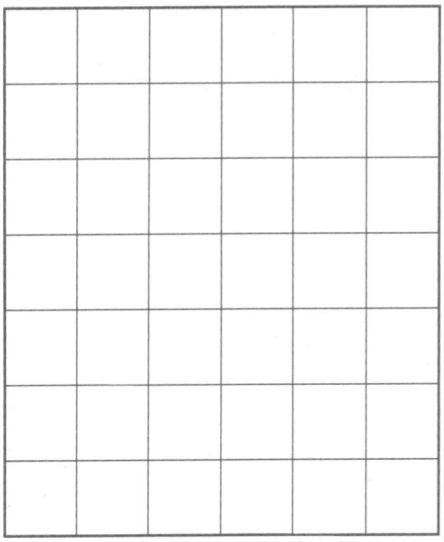

SRB 18 19

5. How many centimeters are in 9.7 meters? Circle the best answer.

A. 907

B. 900.7

C. 970

D. 9,700

SRB 129 315

6. Circle the square that has $\frac{1}{3}$ shaded.

A.

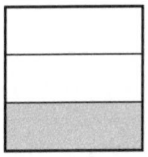

B.

SRB 44

Date Time

LESSON 6•8

Math Boxes

1. Name the ordered number pair for each point plotted on the coordinate grid.

A (_____,_____)

B (_____,_____)

C (_____,_____)

D (_____,_____)

E (_____,_____)

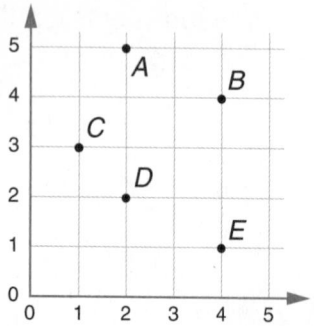

SRB 144

2. Complete the "What's My Rule?" table and state the rule.

Rule

in	out
3.6	2.1
10	8.5
7.2	
	4.9

SRB 162–166

3. $\angle EDF$ is _____ (acute or obtuse).

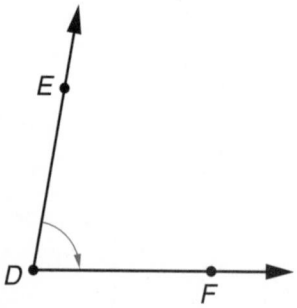

Measure of $\angle EDF$ = _____ °.

SRB 93 142 143

4. Cross off the names that do not belong in the name-collection box below.

32
81 − 49
9 ∗ 4
(5 ∗ 6) + 2
98 ÷ 3
10 + 15 + 7

SRB 149

5. Round to the nearest hundred-thousand.

a. 9,540,234 _____

b. 37,609,034 _____

c. 78,291,554 _____

d. 290,696,332 _____

SRB 182 183

6. Fill in the missing fractions on the number line.

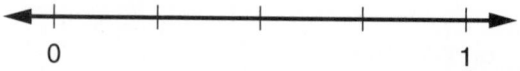

_____ _____ _____

SRB 316

Date Time

LESSON 6•8 A Map of the Island of Ireland

Bantry	B-1	Dublin	F-4	Lahinch	B-4	Omagh	E-7
Belfast	F-7	Dundalk	F-6	Larne	F-7	Tralee	B-2
Carlow	E-3	Galway	C-4	Limerick	C-3	Tuam	C-5
Castlebar	B-6	Gort	C-4	Mullingar	E-5	Westport	B-5
Derry	E-8	Kilkee	B-3	Navan	E-5	Wicklow	F-4

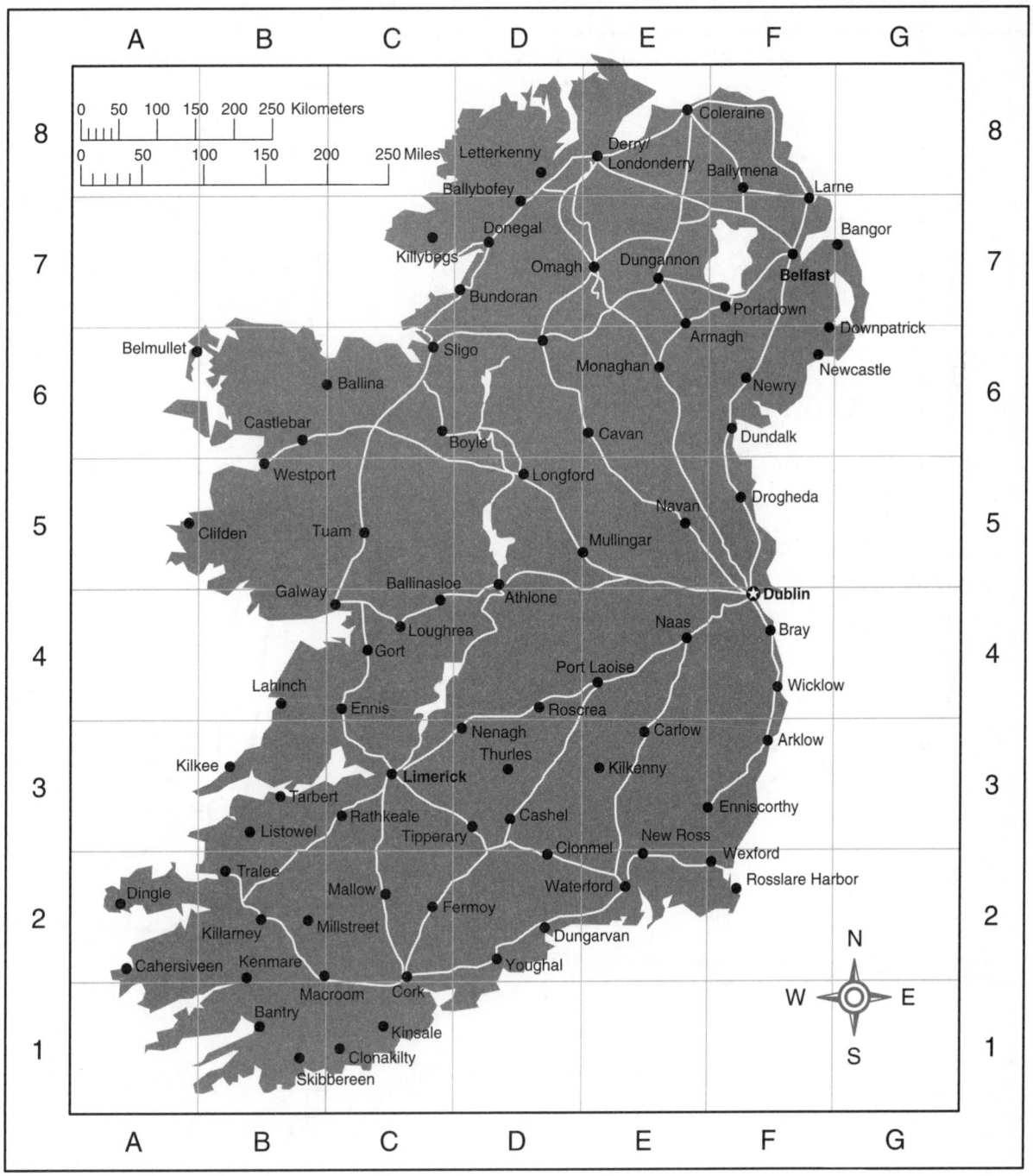

Date Time

LESSON 6•8 A Campground Map

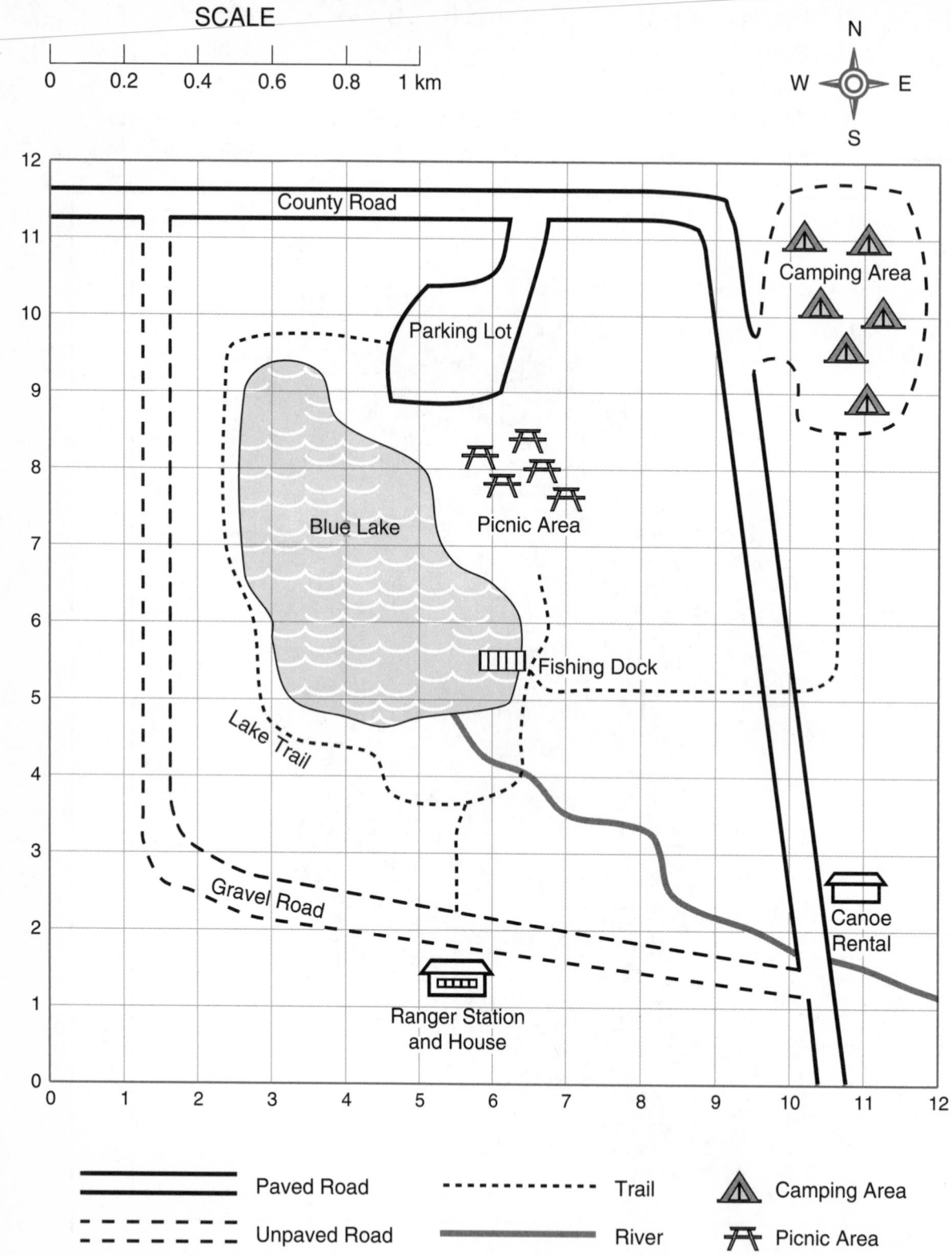

SRB 144 145

Date Time

LESSON 6•8

Finding Distances on a Map

Use the campground map on journal page 162 to complete the following:

1. Suppose you hiked along the lake trail from the fishing dock to the parking lot. Estimate the distance you hiked. About _____ km

2. The ranger made her hourly check. She started at the ranger station. She drove northwest and then north on Gravel Road to County Road. She turned east onto County Road and drove past the parking lot and the camping area. After she passed the canoe rental, she turned right onto Gravel Road and drove back to the ranger station. About what distance did she drive? About _____ km

3. Estimate the distance around Blue Lake. About _____ km

4. You are planning to hike from the camping area to the parking lot. You will stay on the roads or trails. You want to hike at least 5 kilometers.

   a. Plan your route. Then draw it on the map with a colored pencil or crayon.

   b. Estimate the distance. About _____ km

5. Use the ordered number pairs to locate each item on the map. Mark a dot at the location. Next to the dot, write the letter given for the feature.

Campground Features Chart

	Location	Letter
parked car	(5,9)	C
boat	($3\frac{1}{2}$,8)	B
swing set	(8,11)	S
hikers	(10.5,6.5)	H
farmhouse	($\frac{1}{2}$,7)	F

Date Time

LESSON 6•9

Locating Places on Regional Maps

Use the maps on pages 282–293 in the *Student Reference Book* to answer Problems 1–3.

1. Record the continent in which each city is located.

   a. Pretoria, South Africa (Region 1) _____

   b. London, England (Region 2) _____

   c. La Paz, Bolivia (Region 3) _____

   d. Dhaka, Bangladesh (Region 4) _____

   e. Washington, D.C., USA (Region 5) _____

2. Find the approximate latitude and longitude of each city. Record the degrees and circle the correct direction.

   a. Pretoria, South Africa latitude ____ °N or °S; longitude ____ °E or °W

   b. London, England latitude ____ °N or °S; longitude ____ °E or °W

   c. La Paz, Bolivia latitude ____ °N or °S; longitude ____ °E or °W

   d. Dhaka, Bangladesh latitude ____ °N or °S; longitude ____ °E or °W

   e. Washington, D.C., USA latitude ____ °N or °S; longitude ____ °E or °W

3. Each degree of latitude that you travel north or south from the equator is equal to about 70 miles. About how many miles from the equator is each city?

   a. Pretoria, South Africa About _____ miles

   b. London, England About _____ miles

   c. La Paz, Bolivia About _____ miles

   d. Dhaka, Bangladesh About _____ miles

   e. Washington, D.C., USA About _____ miles

LESSON 6•9

Math Boxes

1. Cindy received $40 from her aunt and uncle. She drew a circle graph to show how she will use the money.

   a. How much will she save?

   _____

   b. How much will be spent on clothes?

   _____

   c. On movies?

   _____

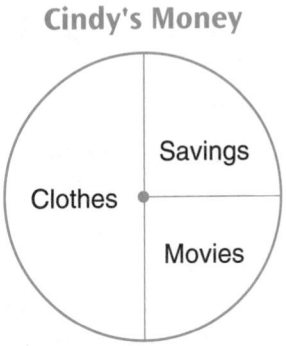

2. Mrs. Moy's students are folding paper cranes for an art project. Each of her 27 students is assigned to make at least 15 paper cranes. What is the least number of cranes the class will have for the project?

   Number model: _____

   Answer: _____ paper cranes

3. Divide with a paper-and-pencil algorithm. Write the remainder as a fraction.

   598 / 3 = _____

4. Which number sentence is true? Circle the best answer.

   A. 33,000,000 < 33,000

   B. 5,200,000 > 9 million

   C. $10^4 = 10{,}000$

   D. six hundred thousand = 10^6

5. For this spinner, which color would you be *least likely* to land on?

   _____

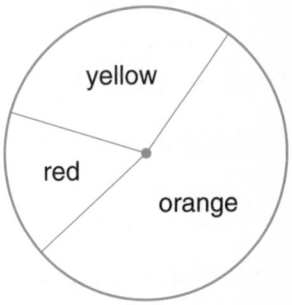

Date Time

LESSON 6•10

Partial-Quotients Division

1. Raul baked 96 cupcakes. He wants to divide them evenly among 3 bake sale tables. How many cupcakes should he put on each table?

   Number model: _____

   Answer: _____ cupcakes

   How many cupcakes will be left over?

   _____ cupcakes

2. The library has boxes to store 132 videotapes. Each box holds 8 tapes. How many boxes will be completely filled?

   Number model: _____

   Answer: _____ boxes

   How many videotapes will be left over?

   _____ videotapes

3. The teacher divided 196 note cards evenly among 14 students. How many note cards did each student get?

   Number model: _____

   Answer: _____ note cards

   How many note cards were left over?

   _____ note cards

4. There are 652 students at a school. The auditorium has rows with 22 seats each. How many rows would be completely filled if all the students attend an assembly?

   Number model: _____

   Answer: _____ rows

   How many students would be left over?

   _____ students

LESSON 6•10

Partial-Quotients Division *continued*

5. 18)864 Answer: _____

6. 509 ÷ 37 = _____

Try This

7. 4,872 / 24 = _____

8. 3,315 ÷ 36 = _____

9. There are _____ players in the league.
(Write a number greater than 300.)

There are _____ players on each team.
(Write a number between 11 and 99.)

How many teams can be made?

Number model: _____

Answer: _____ teams

How many players are left over?

_____ players

LESSON 6•10

Math Boxes

1. Name the ordered number pair for each point plotted on the coordinate grid.

A (_____,_____)

B (_____,_____)

C (_____,_____)

D (_____,_____)

E (_____,_____)

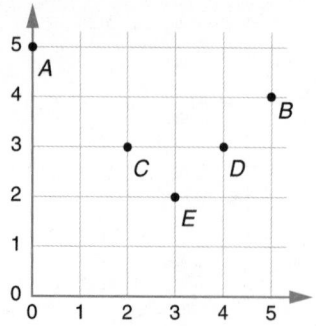

SRB 144

2. Complete the "What's My Rule?" table and state the rule.

Rule

in	out
3.66	7.04
0.42	3.80
8.73	
	12.66

SRB 162–166

3. $\angle NMO$ is _____ (acute or obtuse).

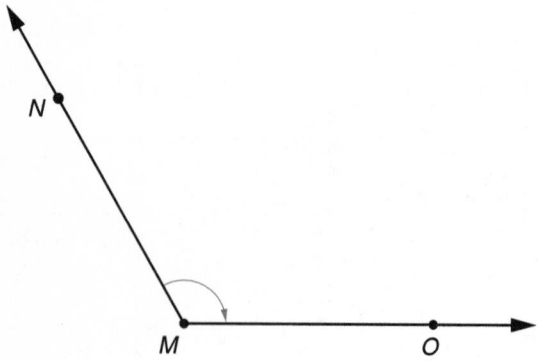

Measure of $\angle NMO$ = _____ °

SRB 93 142 143

4. Cross out the names that do not belong in the name-collection box below.

48
(2 ∗ 3) ∗ 8
100 − 62
18 + 13 + 17
12 ∗ 4
184 ÷ 4

SRB 149

5. Round 451,062 to the nearest thousand. Circle the best answer.

A. 500,000

B. 451,000

C. 451,100

D. 452,000

SRB 182 183

6. Fill in the missing fractions on the number line.

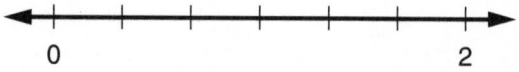

___ ___ ___ ___ ___

SRB 316

Date Time

LESSON 6•11

Math Boxes

1. Fill in the missing fractions on the number lines.

a. $1\frac{2}{5}$ ____ ____ ____ ____ $2\frac{2}{5}$

b. $2\frac{1}{2}$ ____ ____ ____ ____ 5

2. Draw 12 balloons. Circle $\frac{5}{12}$ of the balloons. Mark X on $\frac{1}{4}$ of the balloons.

SRB 44

3. Write five names for $\frac{1}{4}$.

$\frac{1}{4}$

SRB 149

4. a. Shade $\frac{5}{6}$ of the hexagon.

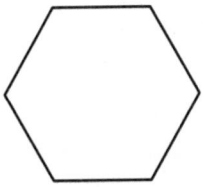

b. Shade $\frac{2}{3}$ of the hexagon.

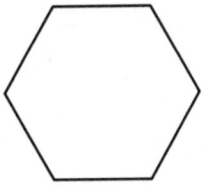

SRB 44

5. Design a spinner such that it is more likely that you will land on red than on green.

SRB 84

Date _____ Time _____

My Route Log

Date	Country	Capital	Air distance from last capital	Total distance traveled so far
	1 U.S.A.	Washington, D.C.		
	2 Egypt	Cairo		
	3			
	4			
	5			
	6			
	7			
	8			
	9			
	10			
	11			
	12			
	13			
	14			
	15			
	16			
	17			
	18			
	19			
	20			

Date Time

Route Map

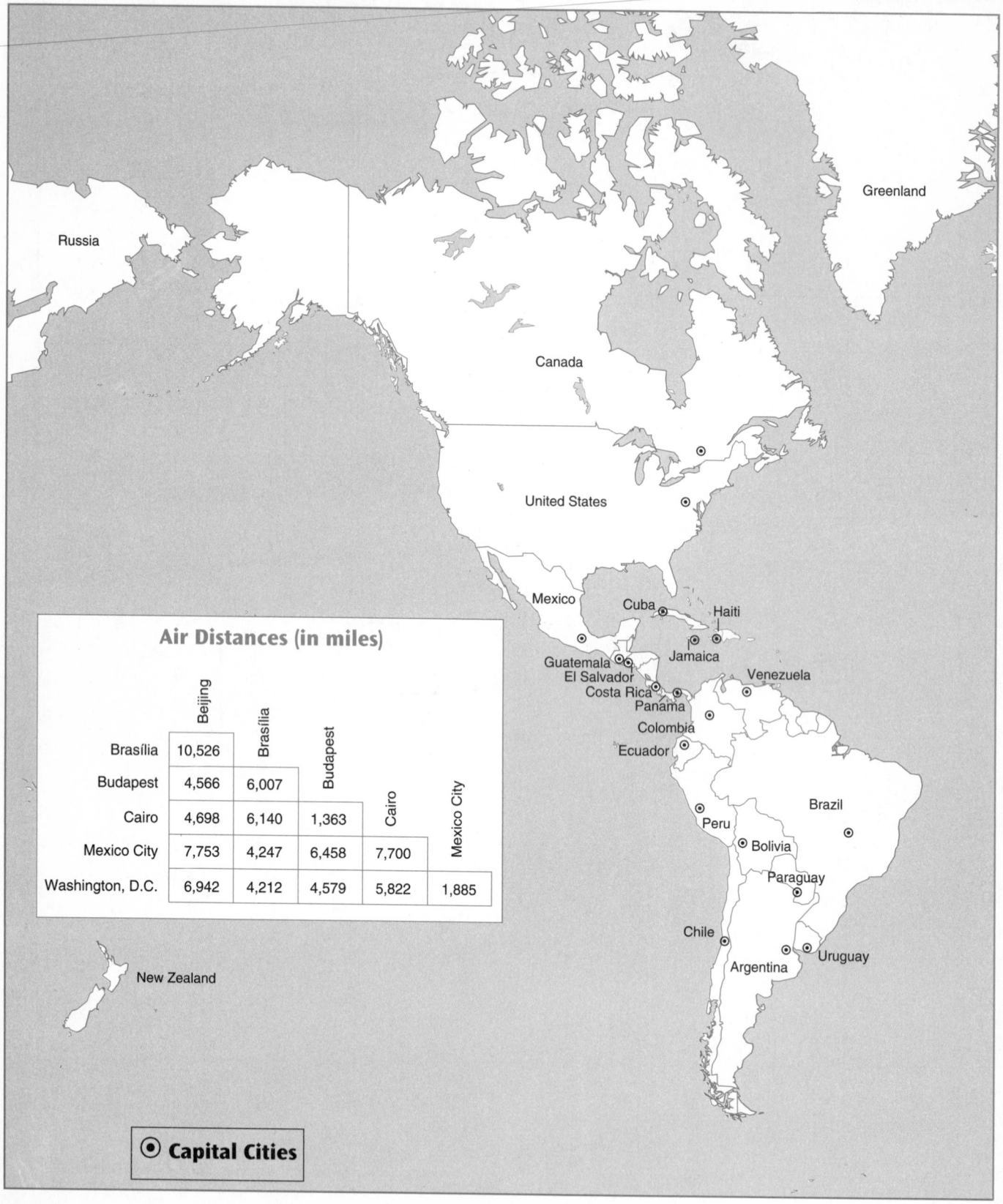

Air Distances (in miles)

	Beijing	Brasília	Budapest	Cairo	Mexico City
Brasília	10,526				
Budapest	4,566	6,007			
Cairo	4,698	6,140	1,363		
Mexico City	7,753	4,247	6,458	7,700	
Washington, D.C.	6,942	4,212	4,579	5,822	1,885

Date Time

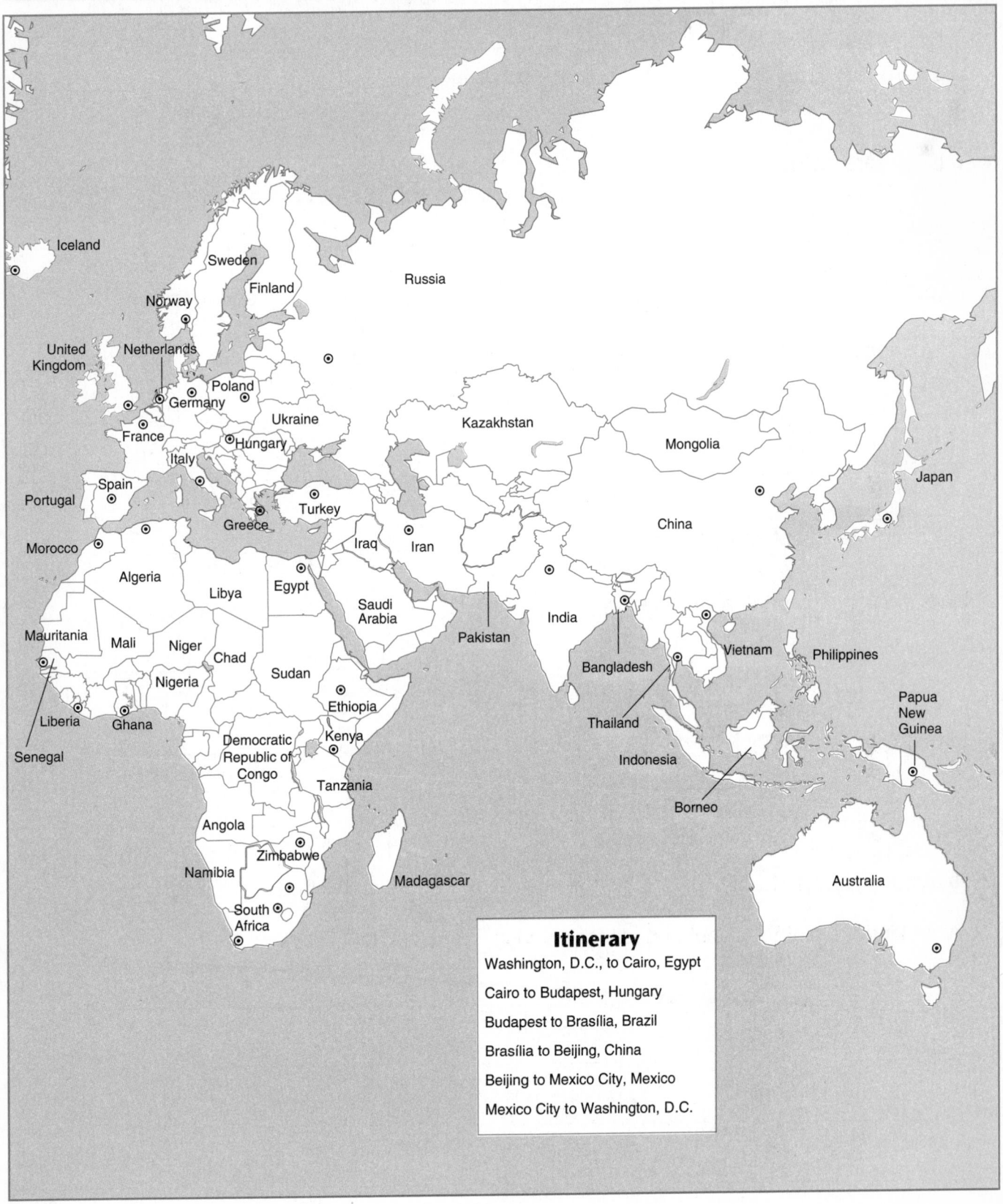
Iceland
Sweden
Finland
Russia
Norway
United Kingdom
Netherlands
Poland
Germany
Ukraine
Kazakhstan
France
Hungary
Mongolia
Italy
Japan
Spain
Portugal
Turkey
Greece
China
Morocco
Iraq
Iran
Algeria
Libya
Egypt
Saudi Arabia
India
Mauritania
Mali
Niger
Chad
Pakistan
Vietnam
Philippines
Bangladesh
Sudan
Nigeria
Ethiopia
Papua New Guinea
Liberia
Ghana
Thailand
Kenya
Senegal
Democratic Republic of Congo
Indonesia
Tanzania
Borneo
Angola
Zimbabwe
Namibia
Madagascar
Australia
South Africa
Itinerary
Washington, D.C., to Cairo, Egypt
Cairo to Budapest, Hungary
Budapest to Brasília, Brazil
Brasília to Beijing, China
Beijing to Mexico City, Mexico
Mexico City to Washington, D.C.

Date _____ Time _____

LESSON 3•6

My Country Notes

A. Facts about the country

_____ is located in _____.
name of country — name of continent

1. It is bordered by _____
countries, bodies of water
_____.

2. Population: _____ Area: _____ square miles

3. Languages spoken: _____
_____

4. Monetary unit: _____

5. Exchange rate (optional): 1 _____ = _____

B. Facts about the capital of the country

_____ Population: _____
name of capital

1. When it is noon in my hometown, it is _____ in _____.
time (A.M. or P.M.?) — name of capital

2. In _____/_____, the average high temperature in _____
month — month — name of capital
is about _____°F. The average low temperature is about _____°F.

3. What kinds of clothes should I pack for my visit to this capital? Why?
_____
_____
_____
_____

LESSON 3•6

My Country Notes *continued*

4. Turn to the Route Map found on journal pages 172 and 173. Draw a line from the last city you visited to the capital of this country.

5. If your class is using the Route Log, record the information on journal page 171 or *Math Masters,* page 421.

6. Can you find any facts on pages 302–305 in your *Student Reference Book* that apply to this country? For example, is one of the 10 tallest mountains in the world located in this country? List all the facts you can find.

_____

_____

_____

_____

_____

C. My impressions about the country

Do you know anyone who has visited or lived in this country? If so, ask that person for an interview. Read about the country's customs and about interesting places to visit there. Use encyclopedias, travel books, the travel section of a newspaper, or library books. Try to get brochures from a travel agent. Then describe below some interesting things you have learned about this country.

_____

_____

_____

_____

_____

_____

_____

_____

Date _____ Time _____

LESSON 4•7

My Country Notes

A. Facts about the country

_____ is located in _____.
name of country — name of continent

1. It is bordered by _____
countries, bodies of water
_____.

2. Population: _____ Area: _____ square miles

3. Languages spoken: _____
_____

4. Monetary unit: _____

5. Exchange rate (optional): 1 _____ = _____

B. Facts about the capital of the country

_____ Population: _____
name of capital

1. When it is noon in my hometown, it is _____ in _____.
time (A.M. or P.M.?) — name of capital

2. In _____/_____, the average high temperature in _____
month — month — name of capital

is about _____°F. The average low temperature is about _____°F.

3. What kinds of clothes should I pack for my visit to this capital? Why?

_____

_____

_____

_____

Date Time

LESSON 4•7

My Country Notes *continued*

4. Turn to the Route Map found on journal pages 172 and 173. Draw a line from the last city you visited to the capital of this country.

5. If your class is using the Route Log, record the information on journal page 171 or *Math Masters,* page 421.

6. Can you find any facts on pages 302–305 in your *Student Reference Book* that apply to this country? For example, is one of the 10 tallest mountains in the world located in this country? List all the facts you can find.

_____

_____

_____

_____

_____

C. My impressions about the country

Do you know anyone who has visited or lived in this country? If so, ask that person for an interview. Read about the country's customs and about interesting places to visit there. Use encyclopedias, travel books, the travel section of a newspaper, or library books. Try to get brochures from a travel agent. Then describe below some interesting things you have learned about this country.

_____

_____

_____

_____

_____

_____

_____

_____

Date _____ Time _____

LESSON 5•11 My Country Notes

A. Facts about the country

_____ is located in _____.
name of country — name of continent

1. It is bordered by _____
countries, bodies of water
_____.

2. Population: _____ Area: _____ square miles

3. Languages spoken: _____
_____

4. Monetary unit: _____

5. Exchange rate (optional): 1 _____ = _____

B. Facts about the capital of the country

_____ Population: _____
name of capital

1. When it is noon in my hometown, it is _____ in _____.
time (A.M. or P.M.?) — name of capital

2. In _____/_____, the average high temperature in _____
month — month — name of capital

is about _____°F. The average low temperature is about _____°F.

3. What kinds of clothes should I pack for my visit to this capital? Why?
_____
_____
_____
_____

LESSON 5•11

My Country Notes *continued*

4. Turn to the Route Map found on journal pages 172 and 173.
   Draw a line from the last city you visited to the capital of this country.

5. If your class is using the Route Log, record the information on journal page 171 or *Math Masters,* page 421.

6. Can you find any facts on pages 302–305 in your *Student Reference Book* that apply to this country? For example, is one of the 10 tallest mountains in the world located in this country? List all the facts you can find.

_____

_____

_____

_____

_____

C. My impressions about the country

Do you know anyone who has visited or lived in this country? If so, ask that person for an interview. Read about the country's customs and about interesting places to visit there. Use encyclopedias, travel books, the travel section of a newspaper, or library books. Try to get brochures from a travel agent. Then describe below some interesting things you have learned about this country.

_____

_____

_____

_____

_____

_____

_____

_____

Date _____ Time _____

LESSON 6•7 My Country Notes

A. Facts about the country

_____ is located in _____.
name of country — name of continent

1. It is bordered by _____
countries, bodies of water
_____.

2. Population: _____ Area: _____ square miles

3. Languages spoken: _____
_____

4. Monetary unit: _____

5. Exchange rate (optional): 1 _____ = _____

B. Facts about the capital of the country

_____ Population: _____
name of capital

1. When it is noon in my hometown, it is _____ in _____.
time (A.M. or P.M.?) — name of capital

2. In _____/_____, the average high temperature in _____
month — month — name of capital

is about _____°F. The average low temperature is about _____°F.

3. What kinds of clothes should I pack for my visit to this capital? Why?

_____

_____

_____

_____

LESSON 6•7

My Country Notes *continued*

4. Turn to the Route Map found on journal pages 172 and 173. Draw a line from the last city you visited to the capital of this country.

5. If your class is using the Route Log, record the information on journal page 171 or *Math Masters,* page 421.

6. Can you find any facts on pages 302–305 in your *Student Reference Book* that apply to this country? For example, is one of the 10 tallest mountains in the world located in this country? List all the facts you can find.

_____

_____

_____

_____

_____

C. My impressions about the country

Do you know anyone who has visited or lived in this country? If so, ask that person for an interview. Read about the country's customs and about interesting places to visit there. Use encyclopedias, travel books, the travel section of a newspaper, or library books. Try to get brochures from a travel agent. Then describe below some interesting things you have learned about this country.

_____

_____

_____

_____

_____

_____

_____

_____